The Green Alternative

Creating an Ecological Future

BRIAN TOKAR

R. & E. Miles
San Pedro • 1987

Peace and Plenty

R. & E. Miles
Post Office Box 1916
San Pedro, California 90733
(213) 833-8856

To my parents
for their patience, love, confidence
and their willingness to be different.

CONTENTS

Introduction—What Does It Mean to Be Green?   1

I.   ORIGINS
   1.   We Are All Part of Nature   9
        Where Did We Go Wrong?   14
        Healing the Wounds   26
   2.   Where Did The Green Movement Come From?   34

II. OUTLOOKS
        What Do Greens Believe In? A Preview   55
   3.   Ecology: The Art of Living on the Earth   58
        Greening Agriculture — A Place to Begin   60
        Water, Air and Forests   64
        Energy, Transportation and Wastes   70
   4.   Social Justice and Responsibility   80
        Work and Technology   86
        Housing, Health, Education and Culture   90
   5.   Democracy in Politics and In the Economy   97
        Democracy in the Green Movement   102
        Creating Economic Democracy   106
        Small is Not Enough   112
   6.   Toward a World of Peace and Nonviolence   115
        What Kind of Defense?   119
        Toward a World Without War   124

III. PROSPECTS
        A Parable   131
   7.   How Can We Create a Green Future?   134
        Greening the Electoral Sphere   141
        Beyond Politics   147

        Notes on Sources   151
        Acknowledgements   169
        Index   171

Introduction:
What Does It Mean to Be Green?

The headlines travelled worldwide in March of 1983 when members of a new political party, known as The Greens, were elected to twenty-seven seats in the West German national parliament. A few weeks later, newspapers featured photos of the new Green legislators in a colorful procession, carrying flowers, tree branches and brightly decorated banners, accompanied by representatives from peace and ecological movements from all across Western Europe. Their march through the streets of Bonn was an exuberant celebration of life. It was a far cry from the staid proceedings that usually mark the beginning of a parliamentary session in Germany or anywhere else.

This was the first time most North Americans had ever heard of the West German Greens, or of the larger Green movement that had by then spread all across Europe. In Belgium, Norway, Britain and the Netherlands, Green or Ecology parties had also begun waging campaigns for local electoral office. With strong ties to local, grass-roots movements for disarmament and ecological awareness, they were helping to bring together people with many different approaches to social change. The Greens were a symbol of hope for a new kind of politics, one in which human and ecological values outweigh the usual demands of power, and working for real peace begins by healing our relationship with the earth.

Over the past several years, groups expressing a Green approach to social change and an ecological way of life have been sprouting up all across North America. Many of these groups can trace their histories back through twenty years or more of popular movements for peace, environmental protection and human dignity.

1

What makes the new Green movement special, however, is its deep understanding of the links between social and ecological problems and the willingness of Greens to question all aspects of our present way of life. The West German experience offers a model of how a sweeping ecological critique of society can appeal to millions of people and help shift the terms of social and political debate on a nationwide scale.

The Greens in West Germany have come to be known by their Four Pillars: ecology, social responsibility, democracy and nonviolence. Greens in the United States have generally expanded this list to include an explicit emphasis on decentralization—the need to reorient both politics and economics toward the local community level. There is often a strong link to the feminist vision of a society that guarantees equal rights to all and embodies the need for personal as well as political transformation. Many Greens also emphasize the search for a new ethical and spiritual orientation, one that reaffirms the place of human cultures within the natural world and seeks to heal the cultural rift between people and the earth that our civilization has imposed.

Guided by these hopes, Greens are involved in a wide range of activities within their own communities. Beginning at the local level, they are working in many ways to begin to bring their ecological visions for society into reality:

—In Vermont, New Hampshire and Maine, Greens have worked with their local Town Meetings to stop the federal government from locating a high-level nuclear waste dump in New England. The Soviet nuclear disaster at Chernobyl sparked a new effort to shut down existing nuclear power plants in the region.

—People in San Francisco are drafting detailed plans for redesigning their city along ecologically responsible lines. Greens in Los Angeles, Berkeley and other cities are also exploring the prospects for ecologically-sound urban living. Such efforts advance the radical notion that cities should invite the influences of the wilder places that surround them instead of trying to pave them over.

—In the Kansas City area, Greens are working to develop ways to bring food directly from nearby farmers to inner city families.

—In Monterey, California, Greens worked with local officials to prevent the open-air testing of genetically-engineered bacteria.

These bacteria, designed to delay late-season frost damage to strawberry plants, could have unpredictable effects on the area's climate and natural vegetation.

—Greens in New Haven, Connecticut, gained widespread recognition by running a slate of Green candidates for city council in a city wide campaign that focused on economic development issues, environmental protection and the effects of poverty and inequality.

—Green activists all across the country have joined in efforts to help stop the forced relocation of thousands of traditional Navajo Indians from Arizona lands coveted for intensive coal mining.

These are but a few examples. Many more will be explored in the following pages.

Another important factor that distinguishes Greens from other political tendencies is the search for an alternative to the economics of growth. Most political leaders in the world today, in the East as well as the West, in the "underdeveloped" South as well as the industrialized North, accept the belief that continued industrial expansion is necessary if we are to feed the hungry and raise people's standard of living. For Greens this is a dangerous myth.

Greens see that unchecked industrial expansion has brought the modern world to the brink of ecological collapse. Our industries release poisons that are altering the world's climate, disrupting food chains, destroying entire forests, and spreading radiation and toxic chemicals throughout the biosphere. New technologies feed a relentless militarism, bringing the threat of nuclear war ever closer, while at the same time virtually enslaving people to machines. Industrial economies create the illusion of affluence for a privileged few, while robbing people around the world of the most basic means of subsistence. It has become clear to many Greens that the hazards of further industrial development now outweigh any possible benefits to humankind.

The Green alternative has at its core a sweeping call for democracy, both political and economic. Democracy, from a Green perspective, does not mean going to the ballot box every few years to choose the people who will make important decisions for you. It means bringing decisions to the local level—to the "grass roots"—where choices can be discussed in an open, face-to-face manner by the people who will be affected by them. In the economic sphere, it means local control over local, small-scale production, oriented toward serving real local needs in a spirit of community self-reliance and careful stewardship of the land. Decentralist economics seeks the enhancement of everybody's quality of life, not the destructive growth for the sake of growth that drives our present system. Federations of ecologically-guided communities can help restore the balances of nature and protect the freedom and dignity of all peoples.

The science of ecology has inspired a new understanding of humanity as one element in an intricate web of relationships that make up the natural world. A close study of nature reveals the profound interdependence of all living beings. The plants, the oceans, the soil and all living creatures are essential parts of a natural, living whole. Removing or damaging one piece of the whole makes life more difficult for all.

The bloodthirsty competition for survival described by thinkers of the 18th and 19th centuries, the proverbial "war of all against all," is a myth created by an increasingly competitive society seeking to justify its heartless ways. It is an image of the world that is not supported by the study of ecology. Natural ecosystems instead reveal a high level of mutual support, reinforcement and interdependence. The most stable, most beneficial habitats are those which are shared· by the greatest variety of species, with the greatest wealth of interrelationships among them. Biotic communities display a tremendous capacity for self-regulation, adaptation and a shared metabolism that preserves the health and stability of the whole. Unity-in-diversity, the sharing of abundance and the elaboration of complexity is the rule; scarcity and competition among species for the control of individual niches are the myths of a competitive society anxious to justify itself.

Ecological diversity has to be valued for its own sake, not just to preserve "resources" as if the whole world is a storehouse of materials for our own future use. At the same time, our respect for human dignity and diversity needs the room to grow and flourish. Nowhere in the rest of nature do we find the pervasive social stratification, the hierarchies of control and domination that plague our modern human societies. Thus an ecological outlook calls into question the very basis of inequality in modern societies and has inspired Greens to seek an end to all relationships of domination. If the patterns of nature favor those species that can complement each other in a mutually balancing way, how can we, with our gifts of consciousness and reason, continue to justify human competition and greed? How can a society founded upon such rigid social hierarchies as is ours—hierarchies based upon race, gender, economic status, language, age, etc.—hope to avoid destroying itself?

Such a thorough restructuring of society will doubtlessly take many generations to fully realize. There is even considerable disagreement among Greens as to just how far we will need to go. But the signs of ecological breakdown and the threat of nuclear annihilation are apparent, and people all over the world are left increasingly powerless to face the consequences. It is both the task and the hope for people living toward the end of the twentieth century to create the foundations for an ecological transformation of society before it is truly too late.

The emerging Green movement, then, has an immense task ahead of it: to illuminate the path toward a cooperative, egalitarian society that can foster the full realization of human abilities in an active partnership with the rest of the natural world. All kinds of people in all spheres of society are beginning to devote their creative energies to this evolutionary process. As in Germany, Greens in North America will need to work to transcend the old political divisions and definitions. It will be necessary to help all kinds of people to express their own needs and desires. This includes conservative farmers struggling to stay on the land, as well as inner-city dwellers just beginning to learn the meaning of self-reliance. A Green movement can embrace displaced factory workers, scientists and inventors, disgruntled secretaries and computer operators, small businesspeople, students, poets, peace activists and all types of nature lovers.

To create a livable future, some deeply held assumptions may have to be questioned: the rights of land ownership, the permanence of institutions, the meaning of progress, the traditional patterns of authority within our society. It will be necessary to explore the real meaning of our traditional values of freedom, equality, democracy and self-reliance. This search will help to reveal ways to begin reshaping our communities and our lives to reflect the kind of world in which people can truly flourish, in celebration of the wonders of nature and the love and sharing of our fellow humans.

The following chapters will attempt to explore what a Green outlook might mean for contemporary life and politics in the United States. The approach will be a personal one, reflecting my own belief, evolved through many years of activism, that a radical restructuring of political and social institutions is necessary for the human species to survive on this earth. Throughout, I will be pointing to various living examples of people working to create a Green alternative. All across this continent, people are living and working today in ways that reflect ecological principles. Some are actively calling themselves Greens; others prefer different labels. Many of the efforts that best reflect Green principles were underway long before there was an organized Green movement in this country. All of their stories are helpful in fleshing out a Green vision. All of my sources are described in the bibliography that follows the final chapter—there, the reader will find the sources of all the quotes I have used, in addition to some pointers that should be helpful in pursuing any of the ideas discussed in these pages.

What follows is a very early attempt to examine the American Green movement in its infancy. Every week, my mailbox brings more ideas and more inspiring examples of people working to create a Green future. I hope the framework I have laid out here continues to prove useful as the movement develops and matures.

I ORIGINS

1.
We Are All Part of Nature

For most of human history, people lived in close harmony with the natural world. Through war and pestilence, the coming and going of empires, epochs of famine and of plenty, seasons of struggle and of celebration, the ancient ties between people and the land prevailed. Through all the upheavals of history and prehistory, nature itself formed the basis for human cooperation and human freedom.

Traditional peoples lived daily with the cycles of nature and the special qualities of the immediate world around them. The natural world was not simply their "environment," but a part of themselves, with the human community a thriving part of the natural whole. The earth was alive and all living beings were her children.

Every aspect of life was a reflection of this intimacy with nature. People gathered food and planted gardens with a deep understanding of the earth's wisdom. They knew how to read her signs and respond to her messages. Their myths, rituals, crafts and even the rhythms of the languages they spoke reflected natural patterns and their daily appreciation of the sanctity and oneness of all life.

And life was everywhere. From the moss-covered boulders to the distant flickering stars, the spirit of aliveness permeated everything. The earth was a goddess, worshipped for her fertility and praised for her abundance. Death, not life, was the great, unexplained mystery of the universe.

Everything seems so different now. Modern civilizations have abandoned the life-affirming qualities of primitive cultures and created a way of life that is increasingly mobilized for death. The vast machinery of our great industrial civilization is dedicated first and foremost to the perfection of ever more sophisticated tools of mass destruction. To fuel our industrial machinery, modern science has reduced all of nature to a set of useful objects that exist to be exploited by us. Nature is seen to be outside of us, an obstacle to be overcome and conquered.

In our time, the earth is being robbed of its ability to recover from the assaults of this civilization. Acid rain is destroying forests in Europe, Asia and North America that have sustained life for eons, while to the south, lush rain forests are being rapidly turned to deserts. Radiation seeps through the air and toxic chemicals ooze out of the ground, spreading cancer and fear through the very water we drink and the food we eat. We are altering the balance between oxygen and carbon dioxide in our atmosphere so drastically that a major shift in climatic patterns is threatened. The "greenhouse effect" promises a short-term warming trend with widespread coastal flooding; a resulting increase in atmospheric moisture toward the poles could accelerate the arrival of the next Ice Age.

Our civilization has evolved a view of the world that we call "scientific," which aspires to understand the fullness of life solely in terms of the physical and chemical properties of inanimate "dead" matter. Death is the norm and life is the puzzle that needs to be "explained." We act as if the world is a machine, just like the machines of our own making. Even the human mind, that seat of consciousness which for millenia was our instrument of attunement to the rest of the natural world, is said to be "explained" by analogy to programmable electronic computers. Our own consciousness thus becomes just another "human resource" to be exploited.

What has humanity gained by adopting this course of development? North Americans tend to believe that they are the ultimate Affluent Society, the freest, healthiest people the world has ever known. But stop for a minute and look around you. The rates of cancer and other chronic diseases continue to soar. More people are going hungry every day. Why? Why have the rates of violent crime, suicide and other signs of social decay reached such unprecedented levels? Why does our government threaten to go to war

every time some small country chooses a course of social development different from our own? And why do we appear to be at a constant state of war against native peoples throughout the world who
simply want to live as their ancestors have for countless
generations?

Before industrial civilization, we are told, people lived in a constant struggle with nature. Basic necessities were frighteningly
scarce and life was monotonous, brutish and short. The development of technology, agriculture and even the rise of large empires
were survival tools of a humanity constantly at war with nature.
This image of "primitive" peoples has been reinforced by historians
and social thinkers from Plato to Marx, and by all of our mass media
from school textbooks to television.

It is a difficult image to shake, but modern archeologists and
those who have gone to study some of the few remaining tribal
peoples are telling us a very different story. They are telling us of
people that have lived for thousands of years in harmony with
nature and with each other. There is evidence, in central Turkey and
elsewhere, of civilizations that lived for centuries with no signs of
warfare. We find accounts of people needing to work less than half-
time to satisfy their basic survival needs, leaving many more hours
and days for creative pursuits, celebration and rest.

People have traditionally worked as much as necessary to live
and eat comfortably, and some primitive societies even stored
surplus food as protection against harsher times. Such food
reserves were not, as is often believed, the innovation of large
empires. Right through the Middle Ages, peasant villages in Europe
were generally free of the pressures to overproduce later imposed
by monarchs and markets, and thus were able to devote nearly as
much time to festivals and religious celebrations as they did to their
own survival needs.

Primitive peoples had a highly evolved understanding of the
world around them and of their place within it. They knew how to
live from the earth without destroying it, whether their food came
by hunting and gathering, fishing, herding or horticulture. In
northern climates, where hunting was generally necessary for
survival, community rituals reflected the sorrow and profundity of
taking another being's life to feed one's people. Each unique cul-

ture had its own ways of expressing their special relationship with the earth and with each other. Let us listen to some Native American voices:

> [Our] cosmology places the Haudenosaunee ["Iroquois"] in a balanced familial relationship with the Universe and the Earth. In our languages, the Earth is our Mother Earth, the sun our Eldest Brother, the moon our Grandmother and so on. It is the belief of our people that all elements of the Natural World were created for the benefit of all living things and that we, as humans, are one of the weakest of the whole Creation, since we are totally dependent on the whole Creation for our survival.
> —Segwalise, introducing the Akwesasne Mohawks'
> *Basic Call to Consciousness*

> My grandma . . . talked to the rocks, the birds, the trees. All creatures, turtles, birds, deer would come to her when she sent out the message for them to come. She did it to show me that you are one, you are relative to everything that flies, crawls, walks, swims, creeps. In that she gave me a sense of comfort in the world, that I needn't fear. All of these things are your relatives and when you are in good relationship then even the animals of the forest will be friends to you and even the things that crawl will show you something wonderful.
> —Dhyani Ywahoo, Tsalagi (Cherokee) spiritual
> teacher, interviewed in *Woman of Power* Number 1

> Ours was a wealthy society. No one suffered from want. All had the right to food, clothing and shelter. All shared in the bounty of the spiritual ceremonies and the Natural World. No one stood in any material relationship of power over anyone else. No one could deny anyone access to the things they needed. All in all, before the colonists came, ours was a beautiful and rewarding Way of Life.
> —Haudenosaunee (Iroquois) *Basic Call to
> Consciousness*

Reverence for nature was usually accompanied by a distinct lack of social stratification. French ethnologist Pierre Clastres has described how tribal systems of governance were carefully designed to prevent the concentration of political power. Especially in the Americas, tribal chiefs had largely ceremonial and diplomatic roles and elaborate political cultures evolved to prevent individuals from exercising coercive authority. Power, to the extent that it existed at all, was generally a matter of personal influence and wisdom, and was rarely able to impose hierarchies of command and obedience on groups of people. Clastres describes one culture in South America where the spirit of mutual aid and interdependence was reinforced by a taboo on eating the meat of an animal one had killed. Each hunter's personal well-being depended entirely on everyone else's fortunes.

Probably the most influential model of democracy in this hemisphere was the Iroquois Confederacy, which was centered in what we now call New York State. Decisions made in the open tribal councils of the Iroquois nations were to reflect the needs of those who would live seven generations hence. The wisdom of women was respected in all decisions, including the decision to go to war. This is a pattern which we find repeated in many cultures around the world:

> There was a time when our lands were torn by conflict and death. There were times when certain individuals attempted to establish themselves as the rulers of the people through exploitation and repression. We emerged from those times to establish a strong democratic and spiritual Way of Life. The [confederacy] of the Haudenosaunee became the embodiment of democratic principles which continue to guide our peoples. The Haudenosaunee became the first "United Nations" established on a firm foundation of peace, harmony and respect.
> —Segwalise

We are now experiencing a time of unusually active appreciation for the contributions of different cultures around the world; at the same time, industrial "development" and commercial agriculture are destroying native peoples and their traditional ways at an unprecedented pace. Many indigenous peoples throughout the world are being coerced, through economic pressure or by military force, to join the larger world economy. From Guatemala to southern Africa to northeastern Arizona, native peoples are promised the "goods" of our civilization and are often forced to accept them at the risk of death or cultural extinction. Our earth may be losing some of its last surviving natural peoples at a time when the rediscovery of more natural ways of life may be essential to our species' very survival.

Obviously, we cannot simply "go back" to a tribal way of life. Even if we could, most people would probably choose not to. People are used to the ways of our own culture and have lost many of the skills and instincts that made the traditional ways possible. We live much closer together and have become used to instant communication and rapid transportation over long distances. People have become thoroughly adapted to a consumer culture that feeds the illusion that our basic life needs will be taken care of for us. The life of the city has provided both an open showcase for world culture and an escape from the provincialism and often rigid social roles of traditional communities.

Tribal cultures are not an ideal to which we should want to return. They are not models for us to copy. Primitive cultures were sometimes quite warlike; some had little respect for the human dignity of people with whom they did not share blood. Sometimes, in times of war or of mounting population pressures, the much-valued tribal democracy would begin to break down. Studying earlier cultures, however, reveals that many different ways of organizing society and relating to nature lie within the realm of human possibility. Understanding our own history as a species is an important tool in the development of a social vision that can, this time, consciously reflect ecological values.

Where Did We Go Wrong?

How did we become so separated from our natural roots? How did we come to see nature as just a storehouse of resources? Social thinkers have pointed to signs of an emerging alienation at many of the important historical crossroads of our species. Let us journey briefly through the history of our civilization and examine some of the key signposts along the road.

For Murray Bookchin, one of our leading social critics writing from an ecological perspective, the process of alienation from nature began with the first signs of domination of one group of people by another. The domination of human by human seems to have led directly to the idea of the domination of nature.

There have always been differences among people, and these differences contribute to the excitement and variety of life. Sometimes, however, as in the tainted history of relations between men and women, these differences come to be exploited to create fixed relations of unequal power. When differences in power are allowed to become embedded in the unwritten social rules of a people, the dominant group creates new social institutions to solidify their power and new justifications for strengthening these institutions.

In his book, *The Ecology of Freedom*, Bookchin shows how the creation of social hierarchies, of fixed imbalances of power and chains of command, marked the transition from societies embedded in nature's ways to those that approached nature with fear and with appeals to supernatural forces. Some authors have suggested that these changes often accompanied transitions from a primarily food-gathering society to one more dependent upon hunting; from a wandering way of life to a more settled pattern based upon agriculture; or from hunting and gathering to the domestication and herding of animals. Others, such as Pierre Clastres, have argued that culture develops independent of such changes in the material base of society. There also appears to be a strong relationship between the rise of hierarchy and the denigration of women's social, economic and religious influence. The elevation of any one group's power over others can seriously compromise patterns of mutual aid and people's shared link to the natural world.

The power of elders to make decisions for a tribe or village may have been the first form of institutionalized power, a power born directly out of the biological insecurity that comes with aging. Institutionalizing the power of elders may have begun to compromise the more fluid, highly personalized types of power that existed in earlier societies. In the earliest times, when a person endowed with leadership qualities passed on, the patterns of influence within the village shifted accordingly. There was no obligation to follow any fixed institutional structure. The earliest gerontocracies, however, began to entrench social patterns of authority.

The first real break may have come as shamans and priests increasingly claimed to mediate the relationship between people and the natural forces. Where priestly power became established, people's direct experience of the spiritual forces inherent in nature was replaced by an abstracted communication through the figurehead of the priest. Priests often banded together to affirm a professional monopoly on spiritual power, enforced by flaunting their skills in manipulating people's deepest psychological fears. Many of the first known cities were founded by such early bands of priests.

Next came the warriors. Warrior clans began as an outgrowth of tribal hunting parties, but increasingly—especially in the emerging river valley settlements of the Middle East, as well as in Japan and other parts of Asia—came to assume the role of independent gangs of "protectors," who would travel from settlement to fragile settlement offering military "services" in exchange for booty and land. Warrior-dominated societies were probably the first to actively accumulate wealth and the first to claim personal ownership of land. On the other hand, the independence of nomadic warrior bands in some parts of the world may have been an important challenge to the consolidation of more rigidly state-like forms.

The development of more stratified societies and ones more separated from nature varied tremendously in different parts of the world. It was likely never a simple linear development, but one that proceeded in fits and starts and which exploited internal distortions and dislocations within tribal societies. Historian Elise Boulding has called the evolution of social hierarchies a "pathological response to stressful conditions." Countless primitive peoples in far corners of the earth were able to resist these patterns until the arrival of the European colonizers, in some cases right up to the twentieth century.

The pattern of emerging social hierarchy that shaped Western civilization was certainly heightened by the harsh living conditions of Near Eastern deserts and river valleys. This life encouraged the development of the first known city-centered cultures, ones that often gained their social cohesion through the manipulation of insecurity and fear. Closeness to the earth and tribal self-reliance gave way to a reliance on centralized institutions and the invocation of supernatural powers, as people sought shelter from the uncertainties of life in such an arid land. It was an age of fortress cities with monumental temple structures and, for many, violence as a way of life. From this Middle Eastern walled "cradle" evolved the European civilization that grew to colonize the entire world.

The first clear glimpse at some of the institutions and patterns of authority and organization that mark our present way of life can be seen in the ancient Egyptian civilization of 5,000 years ago. Lewis Mumford, author of several important works on the history of technology and of cities, has shown how many of the qualities that seem unique to modern industrial society really date back to the Egyptians. In the massive mobilization of human slave labor to build the pyramids, Mumford saw the first example of mechanized production and of a mechanized approach to organizing people.

With only basic hand tools at their disposal, the Egyptians created a human "megamachine"—a massive machine-like mobilization of human, not mechanical, power. They imposed strict divisions of labor, highly standardized tasks and an elaborate system of bureaucratic control. No material need was to be met by this move toward mechanization; necessity was not the mother of invention. The goal was simply the creation of vast monuments to the divinity and immortality claimed by the Egyptian rulers.

This "divinity" was affirmed by the Pharaohs' use of newly acquired knowledge in astronomy and engineering (including the design of elaborate irrigation systems) and was enforced with a ruthlessness unsurpassed in their time. It was the first recorded enlistment of scientific knowledge in the service of social and military power and, through the cult of the Sun God whom the Pharaoh was said to personify, an early use of supernatural beliefs to control human behavior on a grand scale.

Several thousand years were to pass in the evolution of hierarchical, state-centered societies before a written record would begin to outline the ideas underlying the emerging human-nature split. It was the classical Greek philosophers, especially Plato, who first wrote down some of the basic beliefs that underlie our culture's separation from the rest of nature. Our dualistic view of the world, with its rigid distinctions between mind and body, reason and emotion, the real and the ideal, is thought to have been first mapped out in the academies of ancient Athens. Plato proposed a metaphysical world of ideal forms which all true philosophy would contemplate. The world we experience every day is made up of mere copies or shadows of these forms, which therefore are not as worthy of philosophical consideration. Thus our own human bodies, by grounding us in a world of mere appearances, become an obstacle to understanding the true nature of the universe.

The early Roman Christians were strong supporters of this view and institutionalized it through their disavowal of direct spiritual experience. Roman Christianity embraced Plato's anti-naturalism with a vengeance, while at the same time adopting the most life-denying and imperialistic interpretations of traditional Hebrew texts. Images of the death of Christ became the primary ritual objects. All spiritual power was invested in an elaborate Church hierarchy invested with powers of life and death. Nature was endowed as the dominion of Man (that is, the males of our species), the only creature created in the image of the one true God. Counter-currents emphasizing the more humanitarian aspects of Christian teachings have arisen time and again, but the mainstream has invariably reaffirmed its historic role as an ideology of world empire.

Under Christian rule, more traditional forms of spirituality, grounded in direct ritual experience of the powers of the earth, were violently suppressed as "pagan." From the daily brutality of the declining Roman Empire, through the Crusades and the Ages of Exploration and Empire in Western Europe, mass murder in the name of Christianity was the order of the day. The Inquisition of the sixteenth century and the witch burnings of the seventeenth represented efforts to rid Europe itself of such "pagan" practices as folk healing and the free expression of sexuality. The religious zeal that accompanies imperial conquests and colonial wars right up to the present day reflects the same urge of an emotionally and spiritually repressed culture to wipe out those who have held on to more natural ways.

Patterns of economic development and the first conquests of peasant lands by the feudal nobility reflected the same attitudes. When large scale harvesting of firewood came into practice, the sacred groves associated with pre-Christian earth rituals were often levelled first. Ancient shrines were systematically demolished. Historian Frederick Turner describes the consequences:

> In the same way that civilized men (sic) had cleared the earth,
> pruned back the forests, planted villages, towns and cities, so had

> Christianity stripped its world of magic and mystery, and of the
> possibility of spiritual renewal through itself.

He describes the Church as having undertaken a systematic attempt
to remove the sense of divinity from the natural world.

> But its victory here was pyrrhic, for it had rendered its people
> alienated sojourners in a spiritually barren world where the only
> outlet for the urge to life was the restless drive onward . . .

In *Beyond Geography* (from which the above quotes are
borrowed), Turner shows how this drive was to later shape the
European conquest of the Americas. Everywhere the conquerors
established themselves, the natives' traditional ways, as shaped by
thousands of years of intimacy with the land, were replaced by new
ways imported by the invaders from across the ocean. European
land use patterns and European place names were imposed on a
terrain to which they were ill-suited. The fertility of the land and the
once teeming populations of fish and wildlife were all but lost
within a few generations. The native peoples of North America
were nearly wiped out by European diseases to which they had no
natural immunity.

From the European Middle Ages onward, changes in living patterns, the centralization of nation-states and the emergence of new
technologies all helped advance the growing separation of
humanity from nature. For example, the spread of larger-scale agriculture to the heavy clay soils of northern Europe led to the invention of the moldboard plow, which needed to be drawn by large
teams of oxen. With the introduction of such power-intensive
tillage, a cultural shift seems to have occurred. Historians have
noted shifts in local art and mythology from images of a cooperative nature to images of coercion and mastery.

Similarly, the emergence of large scale mining accompanied a
cultural shift toward an objectified view of nature. For centuries, the
practice of mining had been limited by the widespread belief that
extracting ores from the ground was a profound violation of Mother
Earth. This attitude had begun to change by the mid 1500's, when
the first major textbook on mining practices could justify saying,

> "Nature has given the earth . . . to man [sic] that he might cul-
> tivate it and draw out of its caverns metals and other mineral
> products."

Such an attitude would have been virtually unthinkable just a few generations earlier.

These developments evolved gradually, as social and technological changes slowly emerged, evolved together and, in new and unexpected ways, changed people's relationships with each other and with the natural world. But it was not until the 17th and 18th centuries that the idea of dominating nature became the credo of European civilization. This was the period of the so-called Scientific Revolution, a time when the religious dogmas of the Middle Ages would be cast aside, and replaced by a new mechanical view of the universe that established the manipulation and control of nature as the ultimate goal of human reason and human endeavor.

This coincided with the age of the establishment of capitalism as the dominant economic system in many parts of Europe. It was a time of major dislocations of people, especially peasants, many of whom had lived in relatively self-sufficient village economies throughout the Middle Ages. Underwritten by the kings of Europe, the emerging class of merchants and manufacturers was able in many countries to seize peasant lands for commercial agriculture and pasturage and to compel once-independent artisans to bring their skills to jobs in centralized factories. The dehumanizing conditions and long hours of toil in the early factories are now legend. A new philosophy and a new attitude toward technology were necessary to justify these developments.

Many volumes have been written about the Scientific Revolution and the profound changes it brought. To really understand the origins of the mechanical world view that industrial civilization now takes for granted, one should delve into some of the sources listed at the end of this book. For our purposes, it will suffice to just mention a few of the key ideas.

And it is said that nature can be understood only by reduction, that only by reducing her to numbers does she become
clear.

That without mathematics "one wanders in vain through a
dark labyrinth."

It is decided that that which cannot be measured and reduced
to numbers is not real ...
> —Susan Griffin, *Woman and Nature* (with quote from
> Galileo)

In 1605, the great astronomer Johannes Kepler wrote, "My aim is
to show that the celestial machine is to be likened not to a divine
organism but to clockwork." This succinctly sets the tone for the
next three centuries of natural philosophy. Galileo, while formulating the basic principles of gravitation and planetary motion,
asserted that only those aspects of the universe that could be
described using mathematics had any real basis. Mass and motion,
shape, number and force were real; the complex messages of the
senses were illusory and largely irrelevant. This was the Platonic
dualism gone awry.

Descartes and his contemporaries likened the entire universe,
including the human body, to a machine. To Descartes, God had set
the universe in motion according to fixed mathematical laws, just as
the laws through which a king rules his subjects. The purpose of
science was to discover those laws and understand the elements of
nature in sufficient detail so as to "employ these entities for all the
purposes for which they are suited, and so make ourselves masters
and possessors of nature."

It was Francis Bacon, not himself a scientist, who really
popularized the idea of the manipulative power of science. Understanding nature was not enough; "man's" destiny was to literally
enslave nature, to "conquer and subdue her, to shake her to her
foundations." His language reflects the growing reaction against
women during the 17th century and evokes the witchcraft trials still
going on through much of Europe.

The careful systematizing of scientific knowledge in the service of "mankind," especially the ruling elites, was Bacon's lifetime mission. In addition to his scientific writings, he composed a widely-read utopian tract, *The New Atlantis,* describing an opulent ideal society run by an elite of scientist-priests. In Bacon's vision, the "human empire" would not be limited to "power and dominion" over the natural world, but direct intervention in the patterns of nature to "improve" upon it and make it ever-more useful to "màn." He prefigured the attitude of modern genetic engineering that human intervention is necessary to "perfect" nature for our ends. Bacon's own prescription included changing the life cycles of edible plants, the commingling of existing species and the creation of new ones.

These were the ideas that were to shape the course of the Industrial Revolution in Europe and North America, the rise of the factory system and the emergence of modern technology. Every step of the way, people struggled to resist these changes, especially the newly displaced peasants. This resistance took many forms: political uprisings, the destruction of machines and factories and the embracing of new religious and mystical philosophies aimed at rediscovering the aliveness of nature and a personal union with it. The peasant movements that accompanied the great democratic revolutions in England and France were very much inspired by such movements and religious "enthusiasms."

As time went on, however, the coercive power of the new industrial system and increasingly centralized nation-states prevailed. By the middle of the nineteenth century, much of the emerging labor movement had come to embrace industrialism and technological development as vehicles for liberating people from toil. Little did they know that they were only seeing the tip of a technological iceberg.

The emerging socialist movement had its origins in the early part of that century, with the widespread flowering of utopian social visions and experiments that emerged out of the years of the French Revolution. The writings of Marx and Engels, however, labeled industrial capitalism as a necessary stage in the historical process of liberation and thus set the stage for a new kind of political outlook. Their "scientific" socialism sought to give utopianism, with its ethical and communitarian thrust, a bad name. In doing so, they succeeded in bringing the newly accepted capitalist myths of social and technological progress right into the labor movement. The conquest of nature and the primacy of economics over other aspects of social life, ideas which a few generations of capitalists had failed to instill in workers still conscious of their own peasant roots, became cornerstones of socialist thought. Marx and his followers developed a wealth of theoretical tools that to this day help us to better understand the inner workings of capitalism. However, by embracing the factory as a vehicle for liberation, they unwittingly embedded many generations of social activists in the institutions and the overall world view of the dominant social system.

The entrenchment of scientific rationalism and the mechanical view of nature took many forms during the nineteenth century. Physicists like von Helmholtz discovered mathematical laws that appeared to explain our own bodily functions. Darwin asserted, for the first time, that biological evolution was the result of a selection among undirected biological responses to a changing environment. He discarded both religious interpretations and speculations about a life force within nature, opening the door to the modern view of evolution as a product of entirely random genetic mutations. It was the Victorian ideologist Herbert Spencer who turned Darwin's theory into a tale of overt competition for the "survival of the fittest." At a time when industrial capitalism was assuming an increasingly cut-throat quality, with concentrations of private wealth exceeding anything that had been seen before, Spencer's version of evolution was invaluable in helping to rationalize the system's obvious injustices.

Darwin's own analogy between natural selection and the artificial selection performed by animal breeders anticipated a more thorough mobilization of scientific knowledge to manipulate and control nature. By the turn of the century, people were being sold the idea that the chief purpose of human knowledge was to advance industrial technology.

The German chemical industry, in the later nineteenth century, was the first to hire scientists in-house to carry out experiments to aid in the development of new products. By the early twentieth century, a large engineering establishment was in place, especially in Germany and the United States. It was the direct creation of powerful industrial interests, which readily financed its schools and professional societies. The idea of scientific "progress" thus became increasingly tied to the development of new commercial products. The same institutions later encouraged the development of electronics, the computer industry and, with huge infusions of public funds beginning with World War II, nuclear power and the modern high-technology weapons industry.

Along with the rise of electrical and mechanical engineering, a new "social engineering" emerged that enabled industrial managers and social planners to ever-more rigidly bind people to the monotonous rhythms of machines and bureaucracies. In the past few decades, the increasing computerization of all spheres of life has allowed methods of social control and surveillance to evolve to staggering proportions.

> The amount of real leisure a society enjoys tends to be in inverse proportion to the amount of labor-saving machinery it employs.
> —E. F. Schumacher

It has become fashionable in this century to assert that technological development has a life of its own, that technology is out of control and is itself responsible for both the ecological crisis and the social alienation we face. The historical record suggests, however, that our culture's separation from nature is primarily a result of social, not technological changes. The various philosophical developments outlined here were all reflections of social, political and cultural changes which have shaped the way we are taught to view ourselves and the natural world. Many crucial steps have been left out of the discussion in order to focus on the overall picture. But before proceeding to explore ways of healing our relationship with nature, some further attention to the question of technology is necessary.

Technology is neither an independent historical agent, nor is it merely a set of "tools" which can be used to satisfy any social end we desire. Like philosophy, technological development is a product of social realities. But once a particular path of technological development is chosen, that choice can alter the course of social evolution. This is especially true in a society such as ours which so glorifies its technology and which depends upon technological devices and technology-based institutions for its very existence.

Such was not always the case. Bookchin observes that the highly democratic Iroquois confederacy and the brutally authoritarian Inca empire both existed at the same time with virtually the same technological base. Mumford has shown how many of the inventions we now associate with the Industrial Revolution were known centuries earlier, and were used primarily for religious or artistic purposes (or strictly for weapons) long before they were enlisted to do anything "useful."

Glass was used to make ornamental beads thousands of years before it was used to make vessels and another thousand or two before it was used for windows, even though the technical means were available much earlier. The water wheel was first imported from India as a toy and a ceremonial object, and the first windmills and steam pumps were used to work large pipe organs and open and close temple doors. By the 11th century, metallurgy and the use of wind and water power in agriculture and forestry were well-established, but these small steps toward mechanization simply saved labor without significantly altering medieval village life.

It is only in the last century or so that technological changes have had such a striking impact on social development. Today, single technological choices can profoundly shape the course of our future, and the more these choices are left to the "experts" of industry and the military, the less people are able to control their own lives. There are now industrial facilities so large in scale that the relocation of one factory or corporate headquarters can permanently alter the lives of tens of thousands of people.

The feeling that technology is out of control has very deep-lying social roots. The idea of progress that emerged during the nineteenth century has become deeply ingrained in our culture. Our civilization has created a pair of dominant political systems, one ostensibly capitalist and the other claiming to be socialist, that appear to be in competition but which share many basic assumptions about the world. Both systems measure "progress" by the development of technology, reflecting their common belief in the conquest and control of nature. Technologies, under both systems, are chosen on the basis of their usefulness to the military and to industrial megastructures. Both systems thrive on the subservience of people to bureaucracies and an increasing centralization of social and economic power. This is true under both "liberal" and "conservative" governments in the Western "democracies" and during more tolerant, as well as more overtly totalitarian, periods in the Soviet bloc states. In the United States, military technology increasingly determines the course of the technological innovations which allow the economy to continue expanding. Thus the strength of the economy is ever increasingly tied to the expansion of a military-industrial-technological complex. We will return to these ideas in Part II of this book.

Healing the Wounds

In light of these developments, how can we ever hope to heal the rift between humanity and the rest of nature? An ecological awareness calls upon us to free ourselves from the centralization of authority, from technologies of surveillance and control, from economic dependence upon a system that is defiling the earth, justifying mass slaughter and genocide and threatening nuclear annihilation. How can we begin?

A true reconciliation with nature needs to occur on many levels— personal, social, political. A healed relationship with the earth must transcend the artificial boundaries that mark these different spheres, for the attitudes and cultural patterns that sustain our civilization's present ways have been around far longer than most of our present political or social institutions.

Many people are indeed striving to reflect an ecological awareness in their own lives, in an effort to heal their personal ties to the earth. Many people strive to eat healthier foods than they grew up with, trying to grow some of their own where possible. They try to live more in tune with the needs of their bodies and their spiritual selves. They follow the cycles of the seasons, trying to better understand natural patterns and cycles—the sky, the trees, the birds. Many have left the cities to try to create new ways of life—or recreate old ones—in the hills and mountains of this vast continent.

These are some of the more outward signs of a new consciousness that is spreading through our culture, changing it in small and subtle ways that could have a large impact on how we all live in the world. But personal changes, crucial as they may be, are not enough. For our species to survive on this earth, we need to evolve a new culture, a new way of relating to each other and to the world around us. An ecological consciousness has to spread to those who have thus far not been affected by the subtler changes we see occurring. This can create the openings necessary for larger social and cultural changes.

Over the past decade, people in many corners of the continent we call North America have been creating visions and models for a new way of life, grounded in the ways and patterns of nature. These visions have taken many forms and evolved though many separate paths. Over the past couple of years, these visions have begun to come together in ways that can help inform an emerging Green movement.

The most distinctive of these new ecological currents goes by the name of *bioregionalism*. In recent years, bioregionalism has pushed beyond the disciplinary interests of the pioneering geographers and naturalists who first gave it form and become the conceptual basis for a full-fledged social-ecological transformation.

A bioregion is an area of land defined, not by political boundaries—cities, states, countries—but by the natural, biological and geological features that cast the real identity of a place. A bioregion can be identified by its mountain ranges and rivers, its vegetation, weather patterns or soil types, or its patterns of animal habitats, whether birds, ground mammals or humans. The bioregions of North America include such well-known places as the Ozarks, the Sonora Desert, Cascadia (the Pacific Northwest), the Maritimes (northern Maine and eastern Canada) and Appalachia. Bioregionalists further subivide larger regions into individual local watersheds and mountain ranges.

On such a bioregional scale, it becomes possible to more readily determine the most ecologically-sound and sustainable ways for people to live in a particular place. Within a bioregion, people can strive to create self-supporting ways of life that fully complement the flows and cycles of nature that already exist there. The sharing of goods and culture between bioregions then becomes a genuine expression of ecological diversity.

The idea of bioregionalism has evolved gradually in many

different places, but seems to have evolved largely from three major geographic centers. The first is a large region extending from the central California coast up to the Siskiyou Mountains of Oregon. It is a region renowned for its mild, moist climate, its redwood trees and its rich, fertile soils. This bioregion has been the center of a wealth of experiments in earth-centered commmunity living. San Francisco's Planet Drum Foundation has encouraged people to share their experiences and insights about what it means to live in this place. Even within the city of San Francisco, people have begun to explore their ecological history and, through poetry, stories and mapmaking, have reconstructed a picture of how their peninsula looked to its native inhabitants before the days of concrete and sky-scrapers. Throughout northern California, up through the Klamath and Siskiyou mountains that straddle the Oregon border, people have been experimenting with new ways of living on the land. The bioregional poets of this region have inspired people clear across the continent to embrace new (and old) ways of understanding the earth within which we live.

Farther north is the Willamette River Valley region, which includes the city of Portland, Oregon. People affiliated with *Rain* magazine in Portland have been compiling the experiences of people involved with alternative technologies and experimental communities for many years. In 1981 they published their vision for Portland, showing how a regional consciousness can illuminate the solutions to basic urban problems. They showed how people's basic needs could be increasingly satisfied from within the bioregion and how a bioregional economy could better sustain the region's people.

A third major starting point for bioregional activity has been in the Ozark Mountains and plateau, encompassing what we know as Missouri, Arkansas and parts of several bordering states. There, people committed to a bioregionally-focussed way of life have been organizing annual gatherings called Community Congresses. The Congresses bring together people knowledgeable about the essen-tial features of life in the Ozark bioregion—agriculture, forestry, water supplies, health, culture, etc.—and have produced detailed proposals for encouraging the development of each sphere along ecological lines. Between Congresses, task groups continue meeting to work on implementing their proposals. In May of 1984, the Ozark people hosted the first North American Bioregional Congress, at which two hundred people from all across the continent gathered to discuss bioregional ideas and strategies.

In every region where the ideas of bioregionalism have taken hold, it has inspired a new sense of place, of attunement with natural cycles, of appreciation of the common whole of which the human and non-human residents of a region are closely interrelated parts. To Maine's bioregional poet, Gary Lawless, bioregionalism is "an exercise in seeing, learning just where it is that you live, trying to define what 'Home' is all about." To theologian Thomas Berry, "The bioregion is the domestic setting of the community just as the home is the domestic setting of the family." A resolution passed at the first North American Bioregional Congress declared:

> Bioregionalism . . . is taking the time to learn the possibilities of place. It is a mindfulness of local environment, history and community aspirations that leads to a sustainable future. It relies on safe and renewable sources of food and energy. It ensures employment by supplying a rich diversity of services within the community, by recycling our resources and by exchanging prudent surpluses with other regions. Bioregionalism is working to satisfy basic needs locally, such as education, health care and self governance.

Where does your water come from? What kinds of soils are under your feet? What plants are native to where you live? How long is the growing season? Which types of birds can be seen? Where does your garbage go? What myths and legends shaped your region's native inhabitants' view of the land and its cycles? These kinds of questions are of utmost importance to the bioregionalist. David Haenke, a founder of the Ozark Area Community Congresses, has said, "Bioregionalism is rediscovery and reinterpretation . . . of the old ways by those who see we cannot continue in the present profane ways."

Bioregionalism is not just an ethic for rural dwellers. As Planet Drum has shown, an understanding of natural patterns can be cultivated in the city and can shape wiser choices about how everyone lives. They have initiated a Green City project in San Francisco, beginning with a bioregional gathering on the Winter Solstice of

1985 that attracted over 700 people. Since that event, working groups have been meeting to further elaborate bioregional approaches to renewable energy development, agriculture, urban wildlife, transportation and the recycling of wastes. They envision a city of wild-corridor parks and orchards, spacious open plazas and markets, and the enhancement of relationships among neighbors.

A bioregional awareness can help soften the rigid urban/rural dichotomy we now experience, with city dwellers growing more of their own food and cultural events more widely dispersed in the country. A new appreciation of human cultural diversity would begin to replace what Peter Berg, founder of Planet Drum, has termed "the global monoculture." We could begin to create, in the words of the poet Gary Snyder, "the balance of cosmopolitan pluralism and deep local consciousness." For Kirkpatrick Sale, a resident of New York City, bioregionalism means, ". . . the integration into every urban process of a total understanding of ecological principles until the smallest child knows that water does not come from a pipe in the basement and that you can't throw anything away because there is no 'away'."

On the political level, bioregionalism envisions the devaluation of existing political boundaries and the re-emergence of more natural ones. As Gary Snyder explained in a recent *Mother Earth News* interview,

> . . . as an ultimate (and long-range) goal [we would] like to see this continent more sensitively redefined, and the natural regions of North America—or Turtle Island—gradually begin to shape the *political* entities within which we work. It would be a small step toward ecological sanity, and a larger step toward the accomplishment of political decentralization and the *de*construction of America as a superpower into [several] natural nations, none of which would have a budget big enough to support missiles. It would also be a step in the direction of amiable, intimate face-to-face community politics and societies and, ultimately, it would help us develop sane and sustainable economies.

How many bioregions are properly contained in the North American continent? Just how large a region is appropriate for one to focus upon: A local watershed? A mountain range? An entire vegetation zone? These are still open questions. Different types of projects, and different ecological realities, reflect different bio-geographic needs. Natural boundaries are rarely sharply defined and, as ecologists have long known, the most interesting opportunities for interaction and enrichment often occur at the edges.

There are also differences among bioregionalists as to what one means by a "nation." It would be counterproductive to merely reproduce the chauvinism and racism fostered by modern nation-states. Bioregionalists advocate a "breakdown of nations" into self-sustaining, cooperatively-relating, ecologically-scaled entities, which could become models of ecological diversity at the level of human community. One would hope to be able to travel across a mosaic of bioregional communities the way the earliest explorers traveled across Indian villages and tribal territories three hundred years ago—freely and with no apparent fixed borders.

How do we get from here to there, from a culture, technology and social framework thriving on militarism, competition, the despoilation of nature and the exploitation of people to one grounded in ecological harmony and unity-in-diversity? Formidable institutional, political and cultural barriers stand in the way. How can we help each other to free ourselves from all of the economic and psychological dependencies that keep us tied to the existing system? How can we, in the words of Thomas Berry, "join the earth community as participating members?" Peter Berg has written,

> The image of a transformed society isn't difficult to imagine: responsive to the biosphere through use of alternative energy, appropriate technology and sustainable agriculture; smaller political units defined by natural borders rather than straight lines; filling in the qualities of mutual aid, direct democracy, and opportunities for personal creativity and freedom that are nearly absent now.
> —from *CoEvolution Quarterly*, Number 32

For Berg, the first step is to look closely at the broad variety of current examples of people living in tune with the earth and to simply imagine how it could happen wherever you are right now. A truly bioregional effort, he has said, is one that allows the land itself to speak through the medium of human culture. Creative, community-based expressions of bioregional awareness can open the door to a thorough rethinking of present political and social realities.

A second step has been taken by the bioregionalists in the Ozarks: bringing together all of the different sectors of a bioregional community that share an ecological vision. Bioregional Congresses have now occurred in nearly every corner of the United States, from the Pacific Northwest to New York's Hudson Valley. These gatherings are producing detailed ecological visions for several well-defined areas of our continent. They are inspiring people to initiate specific community projects designed to help realize their visions. At a second continental Congress in 1986, a more formal network of bioregional activists was established to encourage the wider sharing of ideas, literature, technical materials and other resources.

The third step is more overtly political, and this is where the Greens come in. It involves closely examining all of the problems our society currently faces and discovering how an ecological understanding could lead to solutions for those problems. A Green outlook addresses global problems such as hunger, the arms race, the poisoning of our air and water and the need to preserve cultures threatened by Western dominance. It requires the empowerment of people in every community in each distinct region to regain control over their own lives. For as long as there are vast inequalities among people, as long as a few powerful interests can freely manipulate the economic and social lives of the many, as long as some nations can threaten others with nuclear annihilation, an ecological society will not be possible.

Greens are creating new models for political and social transformation, beginning at the community level and spanning outward to bioregional, national and international scales. They are involved in community organizing, researching issues, working for ecologically-sound social policies, campaigning for elective office and creating new experiments in earth-centered ways of living and working. Many different forms of political action to protect ecosystems and stop the arms race are being explored. Greens are discovering the interconnections among issues and working to embody a new kind of politics aimed at widespread popular empowerment. Green campaigns for local elective office are geared toward making political institutions subject to more direct popular control at the local level.

The Green movement aspires to be the political expression of a larger effort to reconcile the false separation between humanity and nature. Before exploring the question of how, we will take a brief journey through our recent past and try to discover how Green consciousness has emerged out of the social and political movements of our recent past.

2.
Where Did the Green Movement Come From?

Ask a gathering of European Greens where their movement came from and many of them are sure to point westward across the Atlantic Ocean. For even though the synthesis of political forces that created the Green movement occurred first in Europe, many of the individual political currents brought together by the Greens had their origins here in the United States. To properly understand the origins of both the European and North American Green movements, it is necessary to trace the recent history of peace and environmental activism in the United States and explore our own traditions of local democracy and nonviolent protest.

The real origin of the Green movement is in the great social and political upheavals that swept the United States and the entire Western world during the 1960's. The civil rights, peace and student movements were the first visible signs of the emergence of a new consciousness, soon to be followed by the re-emergence of feminism and of a new environmental awareness. At the same time, a profound cultural reawakening was taking place. By 1966 or 1967, a genuine counter-culture had emerged, which was to shake the foundations of the values Americans had come to take for granted.

The 1960's were a time of widespread cultural, social and artistic experimentation. Materialism, obedience to authority and traditional work and sexual roles were all opened to question. Young people sought liberation from the rigid ways of their parents and of the oppressive institutions of work and school. The cry was to be able to "do your own thing"; the result was a blossoming of new experiments in communal living, liberated personal relationships and spiritual growth. Some experiments floundered and others ended up reproducing the oppressive ways of the larger society merely in a different style. But they were the beginning of an awakening to the idea that the established routines of twentieth century life were not the only way, that it was possible to create new ways of living that could allow people to more fully realize themselves.

The seeds of a new culture were sown during the dark days of the 1950's, when most of America was still aslumber in the air of quietude and conformity that marked the post-World War II years. The scant resistance to the status quo that survived the war was either bought off by the postwar economic boom or rooted out in the Red Scares of the early 1950's. The fifties, far more than the eighties, were a difficult time for anyone who sought to question the existing order.

But something didn't quite fit. Poets from the streets of New York and San Francisco spoke, often deliriously, of the imminent collapse of the staid, button-down ways of white middle-class America. The new rhythms of rock and roll spoke an overt sensuality that respectable parents condemned as dangerous, even subversive. These were the harbingers of things to come.

In the winter of 1955 in Montgomery, Alabama, a black woman named Rosa Parks refused to move to the back of a racially segregated city bus. This simple act of defiance sparked the Montgomery bus boycott which to many marked the beginning of the black Civil Rights movement. A yet-unknown young minister, Martin Luther King, Jr., was enlisted as a spokesperson for the boycott and instantly became a national figure. Marches, sit-ins and large demonstrations spread throughout the South, as blacks increasingly refused to cooperate with the laws that deprived them of their basic rights as human beings.

By the early 1960's, college students and disarmament activists from the North were making regular trips southward to assist black communities in their efforts. They were profoundly changed by the experience. The students, both black and white, returned to their campuses with a personal knowledge of the inequalities and injustices upon which our social system thrives, and with eye-opening experiences of defying that system. They proceeded to turn their knowledge loose on the universities themselves, rebelling against archaic social rules, rigidly controlled courses of study and the suppression of controversial points of view.

Two important currents emerged during the early civil rights and student movements that previously lay buried in conventional accounts of American history. These were the complementary themes of community organization and nonviolent action. Political organization at the community level has been a key undercurrent in this country's history since colonial times; the American Revolution was itself largely born of loosely-knit alliances of local political circles and informal militias. Community-based organizing around economic issues has been important for much of this century. It became particularly widespread in Chicago and other inner cities during the 1940's and fifties, with a wealth of creative nonviolent tactics proving successful against the major urban political machines of the time. During the sixties, many student activists spent their summers in fledgling community organizing efforts in poor neighborhoods of their own cities.

The methods of nonviolent resistance that arose during the Civil Rights movement can also be traced back to the beginnings of American history—many forms of nonviolent noncooperation with the British authorities marked early efforts toward independence. The labor movement and the earliest campaigns for women's rights—the suffragists and their contemporaries—all utilized sit-down strikes, boycotts, human blockades and the like to help advance their aims. Every war this country has ever fought was vigorously opposed in many diverse sectors of society, with people aroused to draft resistance, tax refusal and all means of political action in the name of peace.

The Civil Rights movement, however, was probably the first to raise the theme of nonviolence on a national scale. Images of nonviolent black demonstrators being attacked by white Southern police shook people's confidence in a system of justice so blatantly poised to defend inequality. Participation in these campaigns exposed the brutal underside of American "law and order" for Southern blacks and their supporters alike. By the middle sixties, inner cities across the country were erupting in riotous displays of frustration and rage. Nonviolence had stirred the nation's conscience, paved the way for important social reforms and offered an example for other movements to follow; Black Power, however, threatened the very social stability so valued by middle class America.

The war in Vietnam placed even larger numbers of people of all racial and social backgrounds in direct opposition to the established order. As the government escalated the war, increasing the numbers of military "advisers," bombing North Vietnam, and eventually committing American combat troops by the tens and hundreds of thousands, people's frustration grew. Protest evolved toward open resistance, as draft age men publicly burned their draft cards and many hundreds of people trespassed at the Pentagon and at local military recruiting centers. Hundreds of thousands of people participated in large demonstrations in Washington, D.C. and worked for peace in their own communities. Discontent spread through the ranks of the military, spawning open acts of rebellion against officers and their loathsome orders.

By the early 1970's, the antiwar movement could take credit for substantially slowing the Pentagon's efforts to further escalate the war, discouraging President Johnson from running for re-election and moving all sides, however reluctantly, to the peace table. Opposition to the war forced an early pullout of U.S. troops and may have prevented the use of nuclear weapons against the Vietnamese. The Vietnam War, more than anything else, shaped the experiences and values of those who came of age in the 1960's.

The sixties were a time of change in Europe, too. In May of 1968, French students and workers went on strike together against college rules, workplace regimentation and the stultifying patterns of everyday life. They effectively stopped business as usual in the streets and factories of Paris for the better part of two months. In West Germany, England and other countries, people demonstrated against their own countries' involvements in Vietnam and for social alternatives. To the East, in Czechoslovakia, the movement for greater democracy became so large and vocal that the Russians chose to suppress it with soldiers and tanks. Around the world, people increasingly came to see that sweeping revolutionary changes were necessary to transform a system that thrived on war, poverty and oppression.

The revolution was not to be just political, either. Finding new ways of living, thinking, loving and creating were essential parts of the overall transformation that people were seeking. Though the various political movements and the counter-culture were often in conflict over ideology and personal priorities, they were complementary sides of an overall questioning of values and institutions and a search for new alternatives. In his book, *More Power Than We Know,* nonviolent activist Dave Dellinger described some of the links between political activism and personal transformation that appeared during the sixties:

> In the civil rights and antiwar movements at their best, participants began to rediscover the lost practice of democracy. They began to learn self-reliance, communal solidarity, participatory decision-making, nonelectoral politics, direct action and local, country- and worldwide cooperation. They began to experiment with communal interactions and multidimensional coalition programs that gave satisfaction to the minorities in their ranks as well as the majority. They began not only to savor a reprieve from society's inhibition on love and trust but to explore more dynamic methods of asserting their collective will than pulling a lever on a voting machine every two to four years. . . .

The counter-culture created a sustaining legacy of personal liberation that was able to carry the movement through the continual ebbs and flows of public protest.

Life in "the movement" was not all rosy, however. By 1970, the antiwar movement was badly split over principles and political ideologies. Bitter factional disputes had developed in many areas: differing styles of organization and methods of political analysis; conflicting ways of relating the war to broader social issues; adherence to nonviolence vs. more aggressive forms of confrontation; and revolutionary vs. reformist strategies for long-term change. The more open, democratic ways of earlier years, based on the principle of participatory democracy, gave way in many groups to increasingly authoritarian, even militaristic styles of organization. It was often the women in the movement who reacted the most vocally against these changes.

For years, women had done much of the day-to-day work that kept large movement organizations afloat. With the adoption of more top-down structures and more feverish rhetoric, it became increasingly clear that the antiwar movement was reproducing the role divisions of the larger society—the women did much of the hardest work and the men made all the decisions.

Since the early 1960's, a substantial re-evaluation of women's roles in society had begun to take place. Women were beginning to understand how their restrictive home life, their subordinate economic status and even the most intimate details of their personal relationships were shaped by a male-dominated social system poised to keep women in a subservient role. For women involved in social change work, it was no longer possible to ignore the glaring analogy between the systematic oppression of people of color around the world and the equally systematic oppression of women of all races. It was clear that some major changes were necessary.

Out of this awareness, the Women's Liberation Movement was born. Within its Consciousness Raising groups, women began to see how their most deeply personal experiences of oppression were so often shared with other women. Abolishing war and economic injustice would not be enough to eliminate the oppression women felt in their daily lives, because the personal is also political. To achieve true liberation, many newly-realized feminists argued, it would be necessary to get rid of patriarchy, the social domination by men and, by extension, of all hierarchical relations between people. The prominent radical feminist Mary Daly explains:

> Within patriarchy, power is generally understood as *power over* people, the environment, things. In the rising consciousness of women, power is experienced as *power of presence* to ourselves and to each other, as we affirm our own being against and beyond the alienated identity bestowed upon us within the patriarchy.
> —from *Woman of Power*, Number 1

With these insights also came, for many feminists, a closer identification with nature. For Eco-feminists (a term coined in the later 1970's), the domination of women and the fabled domination of nature by Western Man were born of the same basic denial of the biological nature of human existence, which women had come to personify in the dominant male culture. Eco-feminists set out to reinterpret women's ties to natural cycles as something to celebrate, as the basis for a more thoroughgoing reconciliation of nature and culture and a more profound liberation from socially-imposed roles. For some, this meant founding separate all-women's communities; for others it was a first step toward a healed, nurturing relationship between all of humanity and the rest of nature:

> We know ourselves to be made from this earth. We know this earth is made from our bodies. For we see ourselves. And we are nature. We are nature seeing nature. We are nature with a concept of nature. Nature weeping. Nature speaking of nature to nature.
> —Susan Griffin, *Woman and Nature*

Over the same span of years, a new environmental movement came to the fore in the United States. Early environmentalism developed on a somewhat separate track than the other movements of the sixties. It often had origins in mainstream efforts to conserve natural resources for longer-term use and in the efforts of wealthy elites to keep *their* part of the wilderness free from development and from the intrusions of other people.

By the mid-1960's, however, environmentalism had taken on a new feeling of urgency. Lakes and rivers near large industrial cities were literally dying from uncontrolled dumping of sewage and industrial wastes. They smelled putrid and were unsafe to swim or fish in. City water supplies were threatened. Air pollution had also become a serious health hazard, affecting millions of people in every major city. It was clear that something had to be done.

Environmentalism could not hold on to its exclusiveness and its undertone of elitism for very long. People looked back to the nature writings of the 19th century and discovered that, for people like Thoreau and John Muir, the protection of nature was intimately intertwined with social activism and a critique of industrial society. The science of ecology was also re-examined. In its view of the vital interconnectedness of species in a biotic community, people found a new plea for rethinking the relationships among people in a social community.

[The] integrative, reconstructive aspect of ecology, carried through to all its implications, leads directly into anarchic areas of social thought. For, in the final analysis, it is impossible to achieve a harmonization of man [sic] and nature without creating a human community that lives in a lasting balance with its natural environment.

—Murray Bookchin, "Ecology and Revolutionary Thought," 1965

Most of the environmental laws we take for granted today were passed in the early 1970's. They were the product of overwhelming public pressure on a conservative administration in Washington, D.C., which was simultaneously carrying on an ecocidal, as well as genocidal, war against Vietnam. For some environmentalists, the new laws were almost enough; it was sufficient now to keep close watch on the new environmental agencies and carry on the necessary lawsuits against the most notorious polluters.

But for many, especially people with ties to other social movements, this was not enough. Giving the government more power to regulate was perhaps even part of the problem since regulations were often tailored to the needs of the most powerful corporate interests. Even before the first Earth Day in May of 1970, a few groups were proclaiming ecology as the key to a vision of an entirely new way of life:

> Our cities must be decentralized into communities, or ecocommunities, exquisitely and artfully tailored to the carrying capacity of the ecosystems in which they are located. Our technologies must be readapted and advanced into ecotechnologies, exquisitely and artfully adapted to make use of local energy sources and materials, with minimal or no pollution of the environment. We must recover a new sense of our needs— needs that foster a healthful life and express our individual proclivities, not "needs" dictated by the mass media. We must restore the human scale in our environment and in our social relations....
>
> Ecology Action East Manifesto, 1969

By the middle 1970's the "energy crisis" was upon us. The price of oil nearly quadrupled within a few years. President Carter appeared on TV and said that the effort to stop relying on imported oil was "the moral equivalent of war," a phrase whose irony was lost on many. The nuclear power industry, which had always been highly encouraged and heavily subsidized by the military, claimed to have the solution. By the year 2000, hundreds of nuclear reactors would provide most of our supply of electricity, and this would solve the energy crisis.*

Meanwhile, a quiet opposition to nuclear power development had slowly been spreading across the country. People in rural areas were discovering that they would bear the most immediate consequences of nuclear power's uncertainties. They were supported by growing numbers of scientists, some of whom had once worked for the nuclear industry and had come to realize just how devastating, and how likely, a major nuclear accident would be. But in the atmosphere of the energy crisis, fighting nuclear power through traditional channels, the courts and the regulatory agencies, was proving increasingly frustrating. Stopping a nuclear industry planning to grow six- or eight-fold in a couple of decades was too urgent to leave to judges and government officials.

In early May of 1977, 1,414 nonviolent demonstrators were arrested trying to occupy a nuclear construction site in the small coastal town of Seabrook, New Hampshire. They were seasoned peace workers and young high school activists, teachers, students, farmers and grandmothers. Refusing to pay bail or otherwise compromise with the authorities, people were held for two weeks in National Guard armories all over southern New Hampshire. During their incarceration, a whole new kind of anti-nuclear organization was created, committed to nonviolent direct action and a high level of internal organizational democracy. Over the next few months, anti-nuclear alliances organized along similar lines arose like new spring seedlings all across the United States.

The new generation of activists had learned many of the important lessons of the sixties. Open, consensual democracy was incorporated into the organizational structure of many of the new

*In fact, during the late 1970's only 10% of the oil consumed in this country was being used to make electricity.

Alliances. Feminist values and cooperative group process, open sharing of feelings as well as ideas and strategies, and a strong commitment to internal education helped make these organizations more vital and resilient than many of their counterparts of the Vietnam era.

The apparent urgency of the energy problem helped many people understand that a commitment to ecological values meant creating a new way of living on the earth. The wasteful ways of modern industrial society had to be transformed. In terms of energy use, this meant a shift to the use of more natural and more long-term renewable energy sources—the sun, the wind, water and wood. The movement against nuclear power became closely tied to the Appropriate Technology movement, pioneered by E.F. Schumacher's work in India and by urban "community technology" efforts in New York, Washington, D.C. and the Watts section of Los Angeles.

The concept of ecological living went far beyond just changing energy sources, though. Anti-nuclear activists were often people who had begun to explore the whole spectrum of lifestyle-oriented experiments that flourished in the 1970's—whole foods, holistic health, communal living and a wealth of new approaches to personal and spiritual growth. Large numbers had been influenced by the back-to-the-land movement, through which many people had expressed their disenchantment with the alienating patterns of urban life (including, for some, life in the antiwar movement) by moving to rural areas. There, they sought a more natural and self-reliant existence, which was now directly threatened by the intrusions of the nuclear industry. To the anti-nuclear movement, these re-emerging rural activists brought skills in organic food raising, making clothing and building houses, as well as a wealth of experiences in cooperative living. To the new generation of activists, the back-to-the-landers represented the hope that an ecological way of life was not only necessary, it was indeed possible.

By the early eighties, commercial nuclear power development in the United States had come to a standstill. No new reactors had been ordered for a decade and increasing numbers of planned and even partially built plants were being cancelled. A steadily worsening international political climate moved thousands of anti-nuclear activists to broaden the focus of their concerns to highlight, once again, the problems of peace and disarmament.

In Europe, a similar ecological movement had been developing, but this time the Europeans were several years ahead. One reason is that the European anti-nuclear power movements did not need to "start over" to the degree that their counterparts in this country felt they had to. In many places—Holland, Britain, the cities of West Germany—the movement of the sixties did not appear to collapse toward the end of the Vietnam War as it did here, but continued to evolve and build support throughout the 1970's. Political organizations and counter-cultural institutions could often trace their histories back through several waves of activism and had more readily adapted to the changing political climate. In West Germany, the government's overreaction to the appearance of urban terrorism in the middle 1970's caused thousands of people with activist histories to be banned from government jobs. This made the creation of counter-institutions and an alternative economy much more of a real necessity.

The first nuclear power site occupation took place two years before Seabrook in a small town called Wyhl near the French border of West Germany. Over a two-year period, thousands of anti-nuclear activists had worked closely with local citizens to create a large alternative village on a site that had been chosen for a major nuclear installation. Confrontations with the federal police were sometimes violent but, with visible active support, both local and international, occupiers were able to hold their own until it was clear no nuclear plants would be built in Wyhl.

In other corners of West Germany, direct action campaigns culminating in large site occupations were able to hold nuclear power development somewhat in check. Often these campaigns were accompanied by well-organized Citizen Initiatives that pursued more traditional political channels and involved large numbers of local residents. But in the highly centralized Federal system established in West Germany after World War II, political initiatives at the local level seemed to have little chance of really changing national policies. A long-term campaign involving many thousands of people was not enough to block the construction of a new airport runway though an endangered forest outside Frankfurt. As early as 1978, activists in some regions felt compelled to try running candidates for seats in their state legislatures and in the city council of West Berlin.

In March of 1979, several hundred delegates from a wide variety of ecological, political and alternative-seeking groups gathered in Frankfurt for the founding convention of a new Political Association, The Greens.* The Greens, from the beginning, were to be more than just a political party . They were committed to running candidates for electoral office but, at the same time, were to be an alliance of local, grass-roots activists all across the country, a bridge between the direct action movements, the Citizen Initiatives and people involved in electoral efforts. Electoral campaigns and service in local and national legislatures would serve to "supplement" (their word) local, largely extra-parliamentary efforts.

Greens who ran for elective office would not be the "leaders," nor would they aspire to personal political power. They would be delegates from the various citizens' movements working to advance the ideas of those movements in the country's legislative bodies. The Greens embraced an identity as the "anti-party party":

> We shall not accept any political careerists and we do not want professional politicians. Green state parliament delegates are accountable to their voters. . . . For them it is clear that they will continue working in their local party organizations or citizen action groups even after the election.
> —Program of the Greens of the state of Hesse

Petra Kelly, an American-educated member of the Greens' first parliamentary delegation, explained their approach in her book, *Fighting for Hope:*

> We aim to democratize parliament as much as possible, putting the issues, and the costs of solving them, squarely before the

*The Green organization in West Germany continues to be known simply as *"Die Grünen,"* The Greens. The word "party" has never been a part of their title. The identification of political formations by a color has in fact been a part of the German political tradition for some time—the Social Democrats are the "reds," the Christian Democrats are "black," and so on— but the Greens are the only ones known solely by their color.

public. We must set ourselves uncompromising programmatic objectives in order to stimulate debate and discussion inside and outside parliament. A place in parliament, together with the success of a non-violent opposition movement in the streets, should, we hope, put us in a position to shake people out of their apathy and quiescence.*⁻

The Greens' first national program came to be based upon the "four pillars": the ecological, social, base (or grass-roots) democratic and nonviolent. They proposed a "radical reorganization of our short-sighted economic rationality," a rethinking of the assumptions of industrial growth and the throw-away society. The domination of nature and of human by human would be replaced by an "active partnership with nature and human beings" living in human-scaled and self-governing communities. Both environmental pollution and the miserable living and working conditions many people experience are results of an economy based upon competition and of the "concentration of economic power in [both] state and private-capitalist monopolies." The Greens' social program calls for a new type of social system, one not "ruled by economic power," but governed in an open, participatory manner, in accordance with the principles of ecology and social justice.

Democracy to the German Greens means both the empowerment of people in communities to make the decisions that affect their lives, and the embedding of decentralist principles in the structure of their own organization. Their 1983 program explained:

> A party which did not have this kind of structure would never be in a position to convincingly pursue an ecological policy in the context of parliamentary democracy.

Policy decisions of the Greens are made first at the local level and then are coordinated at large assemblies of delegates where, at least in the early years, policy decisions were made consensually rather than by voting. Green representatives in the parliament are elected along with an alternate so they can return to local political activities

---

**By 1985, Petra Kelly, now an international figure, was to earn the disdain of many Greens for her refusal to rotate out of office after serving for two years (see Chapter 5).

after two years; in most cases, the alternates serve the second half of a four-year term of office. Positions of responsibility within the party are rotated in a similar manner.

> . . . the ecological crisis cannot be solved without putting an
> end to the arms race between East and West, without a new
> economic order between North and South, without social justice
> [and] human emancipation. . . .
> —Rudolph Bahro, East German political exile and
> Green activist, in *From Red to Green*

By the electoral campaign of 1983, it was the Greens' approach to nonviolence and the ending of the arms race that gained them the most national and international attention. The Greens were the only political group in West Germany with a nationwide following that took an uncompromising position against the deployment of American Pershing II and Cruise Missiles across Western Europe. For the Greens, the positioning of U.S. missiles in Europe—and the presence of more Soviet missiles to the East—were new signs of the superpowers' willingness to risk a nuclear war on European soil. To continue to participate in Cold War military alliances was simply to collaborate in the nuclear annihilation of Europe and probably the rest of the world. Opposition to the arms race also meant for the Greens a new kind of partnership with Third World countries, designed to help them become more independent of the U.S.- and Soviet-dominated world economy.

In West Germany and most other West European countries, parliaments are elected by the method of proportional representation. In West Germany, this means that seats in the national parliament (Bundestag) are granted to every political party that gains more than five percent of the national vote, with the proportion of their vote determining how many seats they actually receive. This enabled the Greens to run for office on an undiluted version of their ecological, social, democratic and anti-militarist program. It also empowers elected representatives to adhere to Green ideals, without needing to try to represent those voters who supported other parties with different stances on major national issues. In many ways, the Greens' electoral successes have shifted the focus of political discussion and debate toward a more ecological and non-militaristic vision. At the same time, the Greens have continued to work within the larger West German peace movement for a more grass-roots approach, and have been in the forefront of the search for alternative defense strategies (Chapter 6).

> It has become increasingly important to vote for what one believes to be right on the basis of content rather than wasting one's vote on lesser evils . . . We demand a radical rethink[ing] of all the fundamental issues facing society on the part of the established parties.
> —Petra Kelly, in *Fighting for Hope*

In a parliamentary system, it often takes a coalition of parties to bring together the majority of votes necessary to create stable governments. Since 1984, a major division has erupted in the ranks of the German Greens between those who would seek to join in coalitions with the powerful Social Democratic Party (SPD) in exchange for a more direct role in setting government policies, and those who believe that joining the government would destroy the Greens' ability to function as a principled opposition force of non-politicians advocating thoroughgoing long-term changes. For the "realists" (or Realos, the practitioners of *realpolitik*) changing institutions by gaining political power is the primary goal, even if that means entering into governing coalitions with the Social Democrats. For the "fundamentalists" (or Fundis, those in fundamental opposition to present ways) it is more important to mount a strong ecological opposition to whichever party is currently in power.

In the state of Hesse in 1985, the Greens were offered the opportunity to name one of their own to be Minister of the Environment if they would join in a governing coalition with the Social Democrats. After many months of debate, the Greens agreed to accept the offer, hoping to acquire state aid for a variety of alternative-oriented projects. The new minister, however, was also responsible for administering a number of highly controversial projects, including a reactor that processes plutonium to manufacture nuclear fuel rods, and a proposed new toxic waste depot.

As the 1987 national elections approached, the debate became more strident. The Realos accused the Fundis of sacrificing opportunities to change national policy. The Fundis accused Realos of being willing to compromise away all matters of principle for the sake of electoral success and acceptability. In September of 1986, a national conference of Greens endorsed a proposal to approach the Social Democrats with a coalition offer if the SPD would change its positions on a number of key issues. This move was seen by some observers as the demise of the Fundi position within the Greens. However, Greens across West Germany continue to reject efforts by some Realos to restructure the party along more professional political lines. The 1986 Congress affirmed both the Greens' identity as an instrument of the broader social movements and their critique of the international market economy. When the polls closed in January of 1987, the Greens had won fifteen additional seats and the ruling right-center coalition was left with a considerably narrower majority than before. The continuing debate over "red-green" coalitions returned to the state and local levels, with the Green Environmental Minister in Hesse, Joschka Fischer, resigning in the continuing controversy over the Alkem plutonium plant.

In addition to the forty-two delegates in the national Bundestag, thousands of Greens now sit in state parliaments and in local county and town councils all across the country. Though many Fundis continue to criticize their movement's increasing accommodation to the ways of conventional politics, it is clear that the Greens' parliamentary involvements have changed the way most members view the party's role.*** Indeed, some people who have been active for many years in the West German antinuclear and peace movements have been drifting away from the Greens, believing that their increasing immersion in parliamentary politics has made the Greens just another cog in the machinery of destruction.

***Rudolph Bahro, an advocate of fundamental opposition and one of the most prominent voices of Green politics in Europe, resigned from the West German Green party in June of 1985 after a key parlimentary compromise on the issue of animal experimentation. The Green representatives in the national parliament eschewed demands of animal rights activists for a thoroughgoing ban on animal experiments, opting for a relatively mild form of regulation. For Bahro, this compromise was symbolic of the increasing accommodation of the Greens to the needs of the industrial system, which he sees as a predictable consequence of their increasingly electoral focus.

By the spring of 1984, Green movements and political parties were visible all across Europe. Green delegates had been elected to state local legislatures in nearly every country in Europe and many countries had elected Greens to the European Parliament of the Common Market. In Canada, a Green Party had surfaced with a primarily electoral focus, but most chapters lacked the strong ties to local grass-roots efforts that have characterized many of their European counterparts.

The first move toward a Green organization in the United States occurred at the North American Bioregional Congress in May of 1984. There, a group of people committed to spreading Green ideas in this country met for several days and drafted a pioneering statement of Green principles:

> Recognizing the urgency of our planetary situation and the opportunities for choosing new directions, Green political movements are arising around the world. None of the traditional political forces, whether from left, right or center, is responding adequately to the destruction of ecosystems and the web of crises related to that destruction.
> —Statement of the Green Movement Committee,
> North American Bioregional Congress, May 1984

Later that summer, the Green movement in the United States got off to a somewhat rocky start. Four members of the Green working group at the Bioregional Congress chose to invite a larger group of people to St. Paul, Minnesota, to continue the discussions begun in May. Some believed that a national Green party should be established in the United States within the coming year. Other participants, however, felt that the St. Paul gathering itself violated the principles of openness and grass-roots democracy that have defined the Green movement internationally. Instead of a national body "representing" a spectrum of Green constituencies, many felt strongly that local and regional Green alliances first needed aid in establishing themselves. As a broader Green movement began to develop, regional groups could then federate together through an openly democratic process that would have earned the right to call itself Green.

Thus, the new organization declared itself the Committees of Correspondence, after the locally-based network of people from Maine to Georgia that helped to spark the American Revolution. People from different corners of the country were to return home and begin seeking out people interested in spreading Green ideas. A national clearinghouse was established, which eventually based itself in Kansas City. People inquiring about the new Green network would be put in contact with groups and individuals in their own home regions. By the end of 1986, a regular national newsletter was spreading news among Green groups from coast to coast and several independent Green publications were helping to flesh out the diversity of opinion and organizing methods that give the Green movement its vitality.

The first regional Green conference in New England attracted close to two hundred participants, and twenty local Green groups have since affiliated themselves with the New England Committees of Correspondence. Bioregional groups in northern California, the Midwestern prairie, the Pacific Northwest and dozens of other places have become engaged in discussing Green principles. Green groups in Los Angeles, the San Francisco Bay Area and even New York City have attracted substantial followings. From Maine to California, Greens have undertaken projects in their communities and campaigned for public office on platforms of ecological and social awareness. In July, 1987, the first public gathering of Greens from across the United States attracted over 1500 participants representing more than 80 local and regional Green groups.

In Chapter 7, we will return to the question of how the Green movement can establish itself in this country. We will look at the variety of local efforts that express Green principles and examine some possible next steps. First, though, it is necessary to take a closer look at the principles of Green politics and how they might apply to the pressing ecological and social dilemmas we all face.

II OUTLOOKS

What Do Greens Believe In?
A Preview

The first thing one might notice about Greens in the United States is that there are so many different opinions about just what a Green vision for this country might look like. Greens from different corners of our land, from different social, cultural and economic backgrounds, Greens who live in cities or in rural areas, or whose political outlook has been shaped by different personal experiences might have very different answers to the simple question, "What does it mean to be Green?"

At first, this seems to be quite a dilemma. How can Greens make a real difference if they cannot agree on one unified program for change? The principles of ecology, however, suggest that the movement's diversity should be its greatest strength, not a mark of disunity. The Greens are not a single-issue movement. The goal of reshaping the foundations of this society and its relationship to nature requires that people relish their differences, viewing them as spheres of complementarity rather than as bones of contention.

The Green movement is working to evolve a broad vision for a transformed society that can thrive in harmony with the rest of nature and that fosters harmony, equality and freedom among its citizens. It is a vision firmly rooted in the reality of people's day-to-day experiences in their own communities, but also a global vision that seeks to enhance the web of ecological relationships among all the world's peoples. There are many overriding principles about which Greens agree; the particular interpretations and emphases will necessarily vary widely from place to place and even from project to project.

The Green vision is utopian, not in the negative sense of

appearing impractical or dreamy, but in the reclaimed sense of an unimpaired quest for the kind of society that can truly enhance human possibilities. In the short run, people work for immediate social reforms to alleviate suffering and protect ecosystems, but a Green vision is more than just a list of reforms. It is an effort to understand how all of the pieces of our great post-industrial crisis mesh together as an intricately interlocking puzzle. Then we can work together to discover the new political and social forms that can help us transcend the problems our civilization has created. Understanding the crisis as a whole will help us comprehend the depth and breadth of the changes that are necessary.

Greens are not merely champions of the natural environment. As ecology describes the interconnections among all living things, an ecological politics needs to embrace the interconnectedness of all aspects of our social and political lives and institutions. An ecological sensibility can influence one's view of everything that happens around us and embellish the ways people work together to create change.

The domination of human by human is an ecological problem; the creation of self-reliant, bioregionally-focussed communities is one step toward an ecological solution. In every sphere of life, an earth-centered sensibility can point the way, or more likely toward many mutually supportive ways of transforming social realities. The inspiration comes from a close observation of patterns in the natural world—of complementarity, unity-in-diversity, a fluidity of boundaries and "social" structures and the depth of interconnection among all things. One can begin to map out the paths toward realizing an ecological vision, but one must also be aware that the map will be constantly changing, as new political and social realities open up new possibilities and sometimes close off others.

It would be presumptuous for any single author to attempt to create a Green "program" for the United States. Such a program would be meaningful only after Green activists across the country had fleshed out their own ideas for transforming their home regions along ecological lines. This is a process that has really only just begun. Even then, many would argue that the very idea of a "national" Green program violates the movement's most deeply held principles of decentralism and home rule.

The discussion that follows should be read as a basic outline, a framework, as one person's effort to examine some of our most pressing problems through an ecological magnifying glass and point the way toward some beginning solutions. This is not to imply that the crisis can simply be "solved" within the limitations of the existing system. Rather, it is a way of beginning to comprehend the full depth of the changes that are necessary and of creating new ways to work together.

The discussion here should raise many more questions than it pretends to answer. For every issue discussed, there are equally important ones that are mentioned only in passing, or not at all. I will have succeeded if the chapters that follow help people under-stand what kinds of issues Greens are thinking about and if they stimulate discussions that help enrich the movement's own vision.

For the sake of clarity, my outline follows the West German Greens' "four pillars," the ecological, social, democratic and nonviolent, with some small shifts in emphasis to try to reflect North American political realities and priorities. The categories themselves are not important. The interconnections and points of overlap are much more crucial, as real life always transcends simple categories. For example, where people are going hungry in the world, their plight reflects a failure of industrialized agriculture, of coercive social practices and the myths of economic development. People in many countries have been forced off the most productive food-growing lands by commercial agribusiness, creating profound social disintegration and stripping people of control over the deci-sions that affect their lives. They become dependent upon a faltering urban-oriented international economy and their societies tend to become increasingly regimented and militarized. A similar pattern of interconnecting effects can be drawn for any issue you may choose. As you read the following chapters, pause often to con-sider how the ecological principle of interconnectedness might help illuminate some of the important issues in your own community.

3.
Ecology: The Art of Living on the Earth

Developing a Green perspective on ecological problems is a step toward healing a world that has fallen out of balance. In two centuries of industrial development, this civilization has come to threaten the very survival of the ecological relationships that have evolved on the earth over many millions of years. The most profound changes have occurred in the past thirty to forty years: a visible drop in the number of species of animals and plants, the wholesale destruction of forests, a marked increase in atmospheric carbon dioxide, seasonal "holes" in the ozone layer, the contamination of the atmosphere with radioactive fallout and the proliferation of highly toxic and mutagenic chemicals spreading through our air, soil and water. People knowledgeable about ecology, geology and the chemistry of the earth's atmosphere are beginning to wonder out loud how much further this civilization can go before permanently crippling the earth's ability to sustain life.

All this destruction is in the name of preserving an industrial civilization that allows an elite few to live under an illusion of boundless prosperity and affluence, while people around the world are increasingly deprived of the most basic means of subsistence. We in the "developed" world have built a synthetic shield around ourselves, a shield of concrete, steel and plastic, of chemically treated food and disinfected water, that shelters most of us from the harsh consequences of our own way of life. The increasing squalor of our own inner cities—the homelessness, the pervasive violence and the mounting loss of hope—offers just a glimpse of the system's dark underside. Instead of asking why, Americans continue to promote the short-sighted notion that more industrial growth is the answer to the plight of poor people everywhere.

This mania for growth, in the last thirty years, has spread the damage to far corners of the world previously considered too remote to be useful to us. There is a constant search for more profitable sources of oil and minerals, for more fertile lands to raise our society's staple drugs such as coffee and sugar, and for cheaper labor to run the machines that make our clothes and our computers. Industrialism has spread to the heart of Africa, to the Siberian tundra, to the Amazon jungles and to the frozen Arctic and Antarctic. Meanwhile, the search for more and more people to buy the vast quantities of often worthless goods that the machines produce pulls even more of the world's peoples into our throw-away consumer culture. In the process, vital cultural diversity is sacrificed and Third World countries reproduce the same heightened social divisions that plague our own society.

It is increasingly clear to many Greens that the hazards of an expanding industrial civilization have long outweighed the advantages. Europe's generally higher population densities make the problems appear more immediate there, but the signs of destruction are everywhere.* It is by no means certain that any changes we are able to make as a society will be enough to reverse the ecological and social damage that has been created in our name. People's dependencies on the system might be so deeply ingrained that no amount of political or social reorganization will be enough to reverse the tide. Even a wholesale collapse of the system, of both its Western capitalist and Eastern state socialist branches, might come too late to set back the wave of biological decay that is enveloping our forests, our atmosphere and our own chemical-impregnated bones.

Greens, however, tend to be optimists by nature. For most every cause for hopelessness and despair, one can find a corresponding

---

*By 1987, the persisting radioactive fallout from the nuclear disaster at Chernobyl, the devastating contamination of the Rhine River following a fire at a major Swiss chemical depot and the continuing decline of northern forests due to acid rain led growing numbers of Europeans to believe that the limits of industrialism had come right to their doorstep.

source of hope and encouragement that the necessary reversal of human priorities is beginning. For every outrageous development scheme that goes forward, many more outrageous ones are defeated by mounting political pressure. Every new experiment in ecological living seems to inspire a dozen others. Millions of people are striving to change their own way of life, often in what appear to be very small and personal ways, to become more healthy and more attuned to the earth. We are discovering that every small change can make a crucial difference.

How can our society begin to move more consciously toward an ecological future? The bioregional movement has coined a phrase, "reinhabitation," which means beginning to live more in tune with the natural patterns and cycles of the place one calls home. We need to re-examine some of the most basic features of our lives: the food we eat; the energy we use; how we handle our wastes. We can discover how they have come to reflect our society's alienated ways and how we can create ecological alternatives that can sustain our communities for the long term.

One can begin to press for very specific changes on the local level, which in turn create pressure for larger societal changes. Lasting changes rarely occur by mandate from the top, by changing either the faces or the laws in Washington, D.C. The most important changes originate from the grass-roots, as personal changes are translated into community-based efforts, which can then combine to create irresistible pressure for changes in national policy. In the long run, it is clearly necessary to defuse the authority of decisions mandated from above. Meanwhile, our present national policies are serious obstacles, which can best be overcome by creative initiatives beginning at a local bioregional level.

Greening Agriculture—A Place to Begin

To see how this might work, let us look at the problem of how we feed ourselves, arguably the most vital component of a Green ecological strategy. It has become fashionable in recent years to talk about the "crisis of the family farm," but the crisis goes much deeper than individual farm families being forced off their land. It is a crisis of chemical agriculture itself, and of the commercial food industry, which increasingly threaten the very ability of the land to feed us.

A huge proportion of our food is now produced at huge, heavily mechanized industrial "farms" under the control of a handful of giant agribusiness firms. Their produce is cheap to grow and cheap to buy, but it is increasingly deficient in basic nutrients. It is often trucked thousands of miles to consumers, both urban and rural. Meanwhile, the increasing use of chemical fertilizers, herbicides and insecticides sacrifices the basic fertility of our soils and spreads poison throughout our lands and through the food chain.** Agribusiness' latest plan is to genetically engineer food crops to allow them to grow in increasingly harsh chemical environments and to treat food with high doses of gamma radiation to slow spoilage. Greens in California, New England and other areas are working to expose the hazards of these unproven technologies before they cause irreparable damage to the environment and to human health.

The organic farming movement has blossomed since the early 1960's with a different vision. Once limited to small home-scale growers, painstakingly rediscovering more traditional ways of growing food, the organic movement has proven that even moderately large scale growers can begin to kick their expensive chemical habits. Organic agriculture is devoted to restoring and enhancing natural soil fertility, which is severely depleted by the long-term use of chemical additives. Large field monocultures, which are highly susceptible to disease, are replaced by crop rotation practices that respect the need for soil nourishment and biological diversity. Organic methods result in food that is both more nutritious, and completely lacking in poisonous pesticide and herbicide residues.

A transition to the use of organic farming methods is a key step toward developing local bioregional agricultures. Organic farming is precisely tailored to local climate and soil conditions; it offers

---

**In the Midwest, topsoil is being lost due to erosion at a staggering rate. Nationally, almost five billion tons of topsoil are lost every year and development pressures contribute to the loss of over a million acres of farmland a year. Chemicals and the equipment to apply them account for an ever-increasing share of most farmers' annual costs, with increasing quantities of increasingly exotic pesticides needing to be applied every year.

nutritious food that can satisfy most local needs, while strengthening bioregional awareness. As more people become unwilling to settle for tasteless, overprocessed foods, as more food is grown near large population centers and even within urban neighborhoods, and as the use of energy-efficient greenhouses and non-destructive methods of preserving food encourages more food to be grown in colder climates, a real reversal in agricultural methods could be upon us.

Fundamental changes in attitudes toward agriculturally productive land are also necessary if we are to revitalize local agricultures. First, agricultural land must no longer be bought, sold or taxed as just another marketable commodity. In most places, farmers are now paying taxes on their land based largely on its speculative value to real estate developers. This is one major source of pressure on farmers to sell out and often an insurmountable obstacle to anyone who might wish to begin farming. A few states have begun to implement zoning and taxation methods that assess the value of land according to its agricultural use and create economic and legal obstacles to development. This is one encouraging sign, but stronger measures to prevent the destruction of farmland are clearly needed.

The commercial farm credit system also places an intolerable burden on smaller farms. In order to maintain loan repayments at high interest rates, farmers are compelled to achieve a rate of return that simply exceeds the natural productive limits of the soil. Only those operations that can continue to expand indefinitely or heavily subsidize farm income with funds from other pursuits can thrive under such a system. Food price subsidies have been initiated largely to promote consumer spending in other areas, not to help farmers, and have helped tip the economic balance in favor of the largest-volume producers.

The financial practices and government subsidies that currently favor only the largest agribusiness firms need to be abolished. Public funds can then be used on a local level to help keep arable land open for farming, to help farmers diversify their production and to help existing farms wean themselves from their costly chemical dependencies. Much research is underway on transitional programs from chemical to organic farming that cushion the short-term risk of lost income that discourages struggling farmers from modifying their ways. More such work is needed, along with research efforts to help farmers enhance soil fertility and to elaborate natural methods of warding off harmful insects. New approaches to distributing and marketing food are also needed, based on decentralized, democratically-controlled cooperatives of independent growers. These could be tied directly to the existing networks of consumer food co-ops and to the farmers' markets now established in a growing number of cities. A more decentralized food distrubution system would help relieve the double burden of maldistributed government subsidies and the manipulation of food prices by large commercial distributors and financial institutions, ultimately increasing the availability of fresh, high-quality food.

Many other kinds of experiments are beginning to evolve on a local level. Cities are sponsoring more experiments in community gardening and edible landscaping, everywhere from schoolyards to vacant lots, helping to foster a new appreciation of where food comes from. Land Trusts (see the housing discussion in Chapter 4) are being initiated by towns and cities to remove land from the speculative market. Further changes are on the horizon. Learning how to grow food ecologically could become part of the curriculum in both rural and urban schools. Programs could be developed for passing down farmland from retiring farmers who no longer have children on the farm to young people wishing to begin farming. All efforts to strengthen small and medium sized farms would also help encourage the renewal of rural culture, which is essential for keeping now-threatened farm communities alive and flourishing.

The most important initiatives toward the restructuring of agriculture are beginning to come from farmers themselves. In the summer of 1986, a nationwide United Farmer and Rancher Congress was convened in St. Louis to discuss solutions to the current farm crisis. The Congress, supported by funds raised through the Farm Aid benefit concerts, brought together farmers and farm advocates from all across the country, developed a detailed program of proposed changes in our agricultural system, and is providing farmers' groups with the legal and organizational resources to continue pressing for necessary reforms. Such forums for bringing together farmers as a political force are a crucially important step toward the creation of a new sustainable agriculture.

Water, Air and Forests

A Green strategy of creating changes at the local level can help us move beyond the present political deadlock around many environmental issues. We saw in Chapter 2 how the major environmental initiatives of the early 1970's spurred many important organizations to shift their entire focus toward Washington. In the last ten years, the Washington-based environmental organizations have become increasingly professionalized, increasingly alienated from their constituents and increasingly willing to support political compromises that sacrifice the long-term health of ecosystems in exchange for short-term concessions. Writers such as David Brower, who was pressured to leave the board of the Friends of the Earth organization he founded, and Dave Foreman, who resigned from the staff of the Wilderness Society to found the ecological direct-action group Earth First!, have documented some of the details behind these changes.

The experience of the eighties shows how resistant the federal government and big business can be to environmental lobbying, no matter how well-financed or widely supported it may be. Refocusing efforts around basic ecological issues to a more local level may be the best way for the environmental movement to regain lost momentum. Strong local initiatives to protect watersheds and forests, cleanse our air and water, and protect endangered species could generate the necessary pressure for new national policies as such initiatives and planning processes join together across bioregional lines.

Such efforts can begin to lay the foundations for a new kind of environmental ethic, for a thorough healing of both our attitudes and our own practices toward the natural world. The earth's water, air and forests are resources we borrow from for our survival; it is our responsibility to return them to their natural state. Diversity of species and habitat must be preserved, each for their own sake, whether or not they appear to be useful to human beings. For ultimately, our own survival and that of all other beings depends upon the integrity of the entire ecosphere.

No amount of regulation or de-regulation at the national level can legislate such a sensibility into existence. Environmental initiatives passed in one session of Congress can be readily revoked or weakened by the next unless they are actively supported through local political initiatives. The federal government can strengthen air and water pollution standards; it can also interfere with local efforts to protect the environment, for example by revoking the power of states and localities to ban pesticides or control the nuclear power industry.* As long as we have large concentrations of political and economic power, the interests of one region can be played off against another. Short-term economic benefits can be offered to regions willing to compromise at higher levels of "acceptable" pollution by industries able to relocate their facilities at will.

A narrow focus on Washington actually encourages environmental parochialism, as environmental needs are reduced to political bargaining chips. The continuing controversy over acid rain legislation is a case in point. Midwestern energy producers have successfully fought all efforts to reduce their sulfur dioxide emissions, even though study after study has shown that they are largely responsible for the acid rain that plagues New England, upper New York and eastern Canada. The established political processes allowed acid rain to be dismissed as a northeastern problem until its effects began to show in California and even in some Midwestern lakes. Still, diplomatic pressure from Canada seems to be drawing far more of Washington's attention to the acid rain problem than the actions of environmentalists in the United States. A more popular-based approach, including more focused public education campaigns on the effects of acid rain, efforts to

*In 1986, Congress threatened to prohibit states from banning federally approved pesticides on the grounds that such bans interfered with interstate commerce. Local initiatives to protect people from radiation and toxic chemical hazards have been fought at every turn by federal regulators seeking to protect their comfortable relationships with the industries they are supposed to regulate. When the federal government tried to limit people's access to information on the hazards of chemicals used in their workplaces and communities, people in New Jersey and elsewhere pressured their state governments to challenge these limitations in court. Toxic waste activists, in particular, can tell many tales of local initiatives to protect public health that have been carried out over the objections of federal officials.

mobilize affected groups such as Adirondack fishing enthusiasts and Vermont maple sugarers, and direct pressure tactics such as corporate boycotts and other obstructive actions against the most notorious polluters, might produce better results. Acid rain is an international problem—its solution will require a new sense of foresight and a new ecological ethic that can only emerge by popular initiative and example.

The mounting water shortage in the southwestern United States is another example of an apparently local problem with far-ranging effects. Huge coal strip mines and power plants in the desert, slurry lines running many hundreds of miles, the mining and milling of uranium and an ever-expanding Sun Belt suburban sprawl are draining millions of gallons of water every day from fragile aquifers that are replenished extremely slowly. Farms in California's San Joaquin Valley are being bought up by speculators seeking only to tap them for their water rights. The steady drying up of the lower Colorado River as a result of the over-development of southern California and Arizona has encouraged ever more grandiose water development schemes. One recent proposal calls for a Grand Canal stretching from James Bay in far northern Quebec to maintain water supplies for the Midwest and the Sun Belt. This outlandish scheme, which is being actively supported by Canadian energy and mining interests, would isolate James Bay from the surrounding seas to concentrate fresh water, further despoil lands already threatened by massive hydroelectric developments and create thousands of miles of new canals all across North America, with unknowable ecological consequences. It is another bizarre example of how "local" environmental problems can have unexpected repercussions far away from their source and of the inadequacy of large centralized structures to take such repercussions into account.

Greens are working to evolve a broad new consensus to control development to within natural limits. In recent years, a few cities, including San Francisco, have finally imposed height limits on new buildings and strengthened zoning rules to control the density of future developments. Such efforts are small beginning steps toward controlling pollution, conserving resources and helping to preserve some integrity of urban life. In rural Vermont, where the rapid expansion of ski resorts has put a serious strain on fragile mountain ecosystems, people are now pressing for meaningful development controls. Liability laws governing groundwater have been tightened, the disposal of wastes (treated or not) into rivers and streams has been restricted, and rapidly developing localities are being pressured to plan ahead before their growth runs out of control.

The past decade has brought a massive assault on the integrity of our fragile wilderness areas. The rush to sell off federal lands and to open more land to development has increased the threat already posed by mining, factory-style forestry and more commercialized tourist facilities. The drive to make National Parks "accessible" has brought an invasion of new hotels, condominiums and trailer parks that seriously threaten wildlife habitats. One luxury resort planned for the Yellowstone area would so severely limit the wilderness habitat of the grizzly bear that the bears' very survival is threatened. And during a time when federal budget cuts are depriving millions of basic social services, it is notable that most of these projects could not go forward without federally-subsidized road construction, water diversion and other amenities. The continued over-grazing of millions of acres of federally-owned range lands is similarly sustained by federally funded water and road developments, fencing and herbicide-spraying programs and the programmed slaughter of wild animals found to prey on beef cattle or compete for their food.

The future of the forests is determined increasingly in the boardrooms of large logging, paper and mining companies, where short-term profit is the only important goal. Forests are clear-cut down to stumps and replanted with the single species of tree that companies wish to harvest in the future; "competing" species are poisoned with selective herbicides. The complex web of relationships that forest ecosystems depend upon for their integrity simply does not figure into the cost-benefit analysis.

A Green approach to forestry would place the ecological stability of the forests first, above considerations of "usefulness." Trees should be cut selectively in limited tracts and replanted carefully with an eye to restoring natural diversity. Logging should be prohibited on steep slopes prone to landslides and where soils are highly erodable. Forests should be replanted with a natural diversity of native species of trees, with the rate of future cutting limited by the pace at which the trees are naturally replenished. Greens active around forest issues in northern California have found unexpected allies among lifelong woodcutters who have come to see how accelerated logging by the largest lumber companies threatens the long-term stability of local forest-based economies. Increasingly, environmentally sound forestry practices are seen to be economically as well as ecologically necessary by the people who know the forests best.

Wood is a basic renewable resource, and its use may always be a necessary aspect of ecologically sound living, at least in wooded parts of the country. However, our society's wasteful habits around the use of wood and paper products have to be changed if we are to preserve these resources. A thoroughgoing commitment to the recycling of paper and the elimination of unnecessary packaging are essential first steps.

Ultimately, the restoration of the forests to full health will require not just the acceptance of new conservation measures and an end to acid rain, but a reordering of our social and economic priorities. Political initiatives at all levels are helpful in heading off further environmental damage, but legislated solutions are simply not enough. For the underlying cause of pollution lies not in specific industrial practices, but in Western culture's acceptance of an exploitative relationship to nature for the sake of short-term economic gain. Capitalist economies are not the only culprits, either. The sorry state of Eastern European forests, where the long-term effects of acid rain are rendering untold acres nearly barren of live trees, illustrates how the culture of domination transcends the relatively minor differences between the world's two major growth-oriented economic systems.

No discussion of the dangers of extractive forestry is complete without mentioning the rapid loss of the earth's tropical rain forests. Tropical rain forests are both the most diverse and the most fragile of ecosystems. They are the home of half of the world's animal and plant species, are important in the global regulation of temperature and rainfall patterns and may hold the key to the earth's overall balance of oxygen and carbon dioxide. Almost all of their soil's nutrients lie close to the surface, so when a large area of rain forest vegetation is destroyed, the land can turn to desert in less than five years. If present trends continue, all of the earth's tropical forests will be gone in a few decades as a result of mining activities, tropical wood harvesting, road construction and staggeringly short-sighted efforts to open up rain forest land for cattle grazing—North American fast-food restaurants have been cited as particularly enthusiastic customers for "cheap" beef raised on former rain forest land—and for the resettlement of people driven from densely populated areas. Some ecologists are comparing the large-scale extinction of species that accompanies rain forest destruction to the mass extinctions of 65 million years ago that saw the final demise of the dinosaurs. It is yet another testament to our civilization's tragic short-sightedness.

Along the northern California coast, in the land known to bioregionalists as the nation of Shasta, several community efforts are underway to restore ecological balances once considered lost to the ravages of civilization. The city of Arcata, long renowned as a center of environmental activism, has converted an old town dump on the shores of Humboldt Bay into a lush seventy-five-acre wildlife sanctuary. Where heaps of garbage once marred the landscape, one can now see large flocks of egrets, great blue herons and hundreds of other shore birds. The waters of the Bay are carefully monitored for waste materials that might leach out from underneath the restored marshland soils. A similar project is being envisioned in Berkeley, almost three hundred miles to the south. There, local Greens helped stop a large commercial waterfront development from being built over a former city landfill right on San Francisco Bay. One hillside has since been replanted with a diverse assortment of native shrubs, and plans for a larger urban wilderness area are actively under discussion.

A more ambitious restoration effort, one which is attracting widespread community participation and raising a profound bioregional awareness among local residents, is centered in the Mattole River valley, in southern Humboldt County. The Mattole watershed frames the famous Lost Coast, the most fragile—and arguably the most dramatic—stretch of coast in California. The Mattole shelters one of the last remaining populations of native river salmon in the region, but unsound corporate logging practices over many decades have accelerated erosion, creating levels of river silt that severely threaten the salmon's ability to spawn. In 1980, local bioregionalists began an effort to stop the steady decline of the salmon runs and restore the ecological health of the watershed as a whole.

In an effort that has come to involve scores of local residents from all along the Mattole watershed, groups of people have been capturing fish during their spawning season, fertilizing eggs in controlled environments, raising baby salmon in tanks and releasing them into the river when they appear healthy enough to withstand the existing environmental stresses. Over the next few years, they hope to begin restoring the region's fragile hillsides, rebuilding soils and replanting with native, locally-raised trees. They will be working to slow erosion, thus gradually easing the entire watershed back toward a more natural state. Most of the work is being carried out with volunteer labor and homemade equipment, minimizing the need for outside assistance and the pressures of outside control.

Freeman House, a founder of the Mattole Restoration Council, described the project in the Council's newsletter as:

> . . . the first time that a watershed community of this size has undertaken to understand the workings of their home place, to repair it, and to act as allies of natural systems. Neighbors working with neighbors, rather, than government inspectors crawling over the land. It's an idea exactly as brave, creative and practical as there is community participation in it.

Energy, Transportation and Wastes

The sudden rises in oil prices during the middle 1970's substantially changed Americans' attitudes toward the use of energy. Decades of thoughtless use were called into question by a new awareness of the need to conserve. Homes and public buildings began to be designed with a view toward saving energy. Many people began choosing their automobiles on the basis of gas mileage. The demand for electricity began to level off and plans for over a hundred nuclear power plants were scrapped. Solar and wind energy became acknowledged realities. By the middle 1980's, not only had the energy shortages once forecasted by industry pundits been averted, but there was a visible energy glut that seriously threatened the economies of oil-producing countries.

We still have a long way to go, however. The structure of the utility industry still actively encourages waste, even while officials mouth the rhetoric of conservation. Excessive investments in unneeded nuclear plants throughout the 1970's are now causing electric rates to double in many areas of the country. Nearly everyone has seen working examples of solar-heated and super-insulated buildings, but devices for harnessing renewable energy are still far too expensive for most people. Meanwhile, the nuclear industry continues to receive billions of dollars a year in federal subsidies for research, waste handling and artificially reduced insurance liability.

The nuclear meltdown at Chernobyl in April of 1986 is now thought to have released more long-lasting radiation into the environment than all previous nuclear incidents combined, including the fallout from every nuclear weapon ever exploded. Over a hundred thousand people in the Soviet Union lost their homes. People across Europe had to destroy thousands of acres of topsoil and abandon major food supplies due to radioactive contamination. In the far north, many native people are no longer able to herd reindeer for food, due to the long-term contamination of the native mosses and lichens upon which the animals feed. It is only a matter of time before an accident of similar proportions occurs here if we do not begin immediate shut-down of our country's own nuclear industry and a serious pursuit of alternatives.

Communities can begin taking their energy futures into their own hands by municipalizing local facilities for generating and transmitting electricity. This is particularly feasible in cases where local private utilities are faltering under the weight of past nuclear investments. These new municipal utilities should be demo-cratically controlled by local residents, unlike the many rural electric "cooperatives" which have fallen into the hands of commercially-minded utility officials. Such utilities, relieved of the pressure to generate ever more energy to sustain high profits, could begin to seriously undertake conservation measures. They could offer long-term interest-free loans for thoroughly weatherizing homes and installing solar equipment. They could help finance small local wind power stations and restore inactive hydroelectric dams. They could offer local industries real incentives to reduce energy use though the cogeneration of electricity from waste steam and other innovative measures.

Such measures, however, only begin to address the vast excess of energy consumed daily by our modern industrial megamachine. Per capita energy consumption in the United States is two and one-half times the European average and thousands of times that of many Third World countries. An ecological future demands that people strive to drastically change this country's energy use patterns by limiting both waste and overproduction, by recycling materials and by limiting the use of electricity to the purposes for which it is really needed. For example, the inefficient use of electricity for space and water heating makes for cheaper home construction in the short run, but is vastly wasteful of both money and resources in the long run.

Transportation accounts for a huge proportion of the energy wasted in our country, and an even larger proportion of urban air pollution. Substantial improvements in automobile efficiency have occurred in the past fifteen years and catalytic converters for partially controlling noxious exhaust fumes have become the norm. However, most urban centers are still on the verge of suffocation from both the smog and the physical congestion caused by excessive automobile traffic.

Public transportation in this country has undergone a steady decline since the late 1920's when General Motors began buying up local trolley services and replacing them with fleets of diesel-powered buses. The shift in funding priorities from rail transportation to road construction continues to this day. We have entered the age when our railroads are in such a severe state of neglect that it is sometimes cheaper to fly between two cities than to take a train. In most suburban and rural areas, private automobiles are the only remaining means of transportation.

The Greens in West Germany have called for a complete ban on the construction of new highways, as well as new airports and other huge anti-ecological projects. Such a measure here would be a useful first step toward renewing both our transportation system and our economy. The money that is saved could go toward revitalizing urban subway systems and beginning to restore the networks of trolley lines that once roamed the countryside. It is an absurdity that subsidized rail transportation is such a matter of controversy in this country, while highway subsidies continue unabated. This is in contrast with most of Europe, where comfortable, publicly supported long-distance railroads have been looked upon with pride for a hundred years.

The Europeans are also far ahead in the area of bicycle transportation. In many European cities, commuting by bicycle is commonplace and roadway patterns have been adjusted to make it easier to commute to work on two wheels. Bicycle commuters in most American cities risk life and limb at every turn. A few Western cities (Palo Alto and Davis, California and Eugene, Oregon, for example) have redirected traffic to accommodate bicycles and have built bicycle-sized bridges and underpasses to avoid dangerous intersections. However, the development of bicycle paths for either commuting or recreation has yet to become a priority in the vast majority of our cities.

During the 1970's hopes were raised that a significant portion of our fossil fuels, including gasoline, would soon be replaced by easily renewable biological fuels, such as alcohol from fermented farm wastes and methane from bacterially digested sewage. Yet these alternatives are still hardly available, while billions of dollars in public funds have been pumped into exotic and destructive synthetic fuel technologies. Farmers in some areas have learned to fuel their own farms with methane digested from animal manures. Such methods, applied to landfills and municipal sewage, could contribute substantially to solving both our energy and waste disposal problems.

If the experience of the 1970's began to deflate the myth of an unlimited energy supply, the 1980's could be remembered as the decade when many Americans began to see through the myth of the disposable society. Nature is beginning to show us that we can never really throw anything away, as communities all across the country face serious waste disposal crises. Town dumps and landfills once treated as though they were bottomless are contaminating water supplies and befouling deep underground aquifers. Barrels of toxic industrial wastes are leaching cancer-causing chemicals into soils and streams, and even ordinary household wastes have become a visible environmental problem.

Greens from all over the country have become involved in efforts to find new approaches to the problem of waste disposal. They have been spurred on by local officials seeking to build large new incinerators for the disposal of wastes. Such incinerators can produce a small amount of steam or electricity from the wastes they burn and are being misleadingly sold to municipalities as a combined solution to both their waste and energy problems.

Activists have discovered that such facilities produce dioxins, the deadly carcinogens which were first discovered as impurities in potent herbicides such as the Agent Orange used to defoliate Vietnam. Large amounts of toxic heavy metals and acid gases are also emitted by even the most advanced incinerators. Even if these emission problems could be solved, however, incinerators represent an expensive, short-sighted approach that encourages people's wasteful habits and delays the search for a lasting solution.

In many places, people are taking the longer view and have begun to devote local resources to reducing, reusing and recycling wastes. Communities in New York state, California, New England and elsewhere have established goals of reducing the amount of waste they must dispose of by 60–80%, combining a variety of old and new conservation methods. Paper products are the most easily recycled, facilitating the production of everything from newsprint to building insulation. Glass and metals are readily melted down for re-use. Brush, wood products and food wastes could fuel large and small municipal composting facilities, which can now produce valuable soil enrichments in a matter of days. Municipal-scale composting was practiced throughout the American Midwest in the 1950's and is increasingly common in Europe and Japan today. New technologies have become available for recycling everything from automobile tires to plastic containers.

For a fraction of the cost of building new landfills or incinerators, towns, cities and federations of towns are building facilities that encourage recycling. New collection systems for home-separated trash (with appropriate education and outreach programs), plants to manufacture cellulose insulation from newspapers, and facilities for separating and processing glasses and metals (partially processed recyclables are more readily sold) can be designed to an appropriate scale for local or regional needs. A company called Urban Ore in Berkeley, California, has been successfully "mining" local landfills for building materials and other goods that can be reused, as well as reclaiming appliances and machinery that can be repaired and sold. This helps free up scarce landfill space for the small proportion of materials that are not presently recyclable. The cost of developing such operations is readily recovered through savings in future disposal costs, costs which have been capturing an ever larger share of many local budgets. At the same time, it represents a meaningful step in transforming our wasteful society to one grounded in conservation.

The disposal of toxic industrial wastes represents a more complex political problem, one that was not widely acknowledged until the Love Canal disaster made national headlines in the late 1970's. Of all the ecological disruptions people face, toxic waste presents the most immediate danger to public health and well-being. Our society is ever more dependent on chemical and industrial products whose manufacture involves the use of highly dangerous materials. The manufacture of chemicals, the chrome plating of metals, the use and disposal of many common industrial solvents—all these processes expose both workers and communities to toxic hazards. These hazards may be ignored for years, until a disaster like Love Canal or Bhopal makes the news, or when corroded waste barrels are discovered near a local stream or roadside.

The federal government and over thirty states have passed Superfund laws, through which fees collected from toxic waste producers are pooled to clean up hazardous sites. But by 1986 only six sites out of many thousands had been treated under the federal Superfund. Tremendous public pressure led to a strengthening of the federal Superfund in 1986, despite strong opposition from the chemical industry and the Reagan administration; however, people living in the vicinity of toxic disposal sites are still suffering the greatest consequences.

Toxic waste activists have become increasingly aware that it is not enough to simply clean up one's own backyard. When toxic barrels and contaminated soils are removed from one site, they are most often shipped to a larger hazardous waste dump in another community. Industries must instead be pressured to drastically reduce toxic chemical use at the source. Methods for replacing toxic chemicals by safer materials, and for recycling metals and solvents, are readily available, but do not bring the short-term economic savings that most corporations require for investments in new processes. A strengthened system of both regulations and incentives is necessary to reduce toxic chemical use in short order, especially for the large industrial users that represent the greatest share of toxic chemical production and handling. We can no longer keep sacrificing public health—nor the health of workers in chemical-dependent industries—for the short-term gain of corporate polluters.

Like the energy crisis of the 1970's, the present controversies over waste disposal are opportunities to raise the question of what we produce as a society and how we define our needs. Our economic system floats adrift on a silver-lined cloud promising unlimited material goods to satisfy limitless needs. It is equally founded on the myth of disposability. International trade and credit systems thrive on the destabilizing and ecologically destructive myth of ever-increasing production.

As a society, we need to raise the question: "What do we really need to live a satisfying life?" If something cannot be manufactured or built or grown without causing irreparable ecological damage, can't we strive to create something to take its place, or simply decide to do without it? Nobody's personal economic well-being should depend upon depriving other living beings of their ability to be healthy and thrive. It does not matter whether those beings are fish at the bottom of a lake, neighbors on the other side of town, workers in a chemical plant, or people of other cultures in other parts of the world. Ecology teaches that everything alive is part of an intricately interdependent whole. Every assault on the well-being of any piece of the whole will come back to haunt us in new and unexpected ways.

Many sectors of the environmental movement have embraced the idea of the "information society" as a way to check the excesses of industrialism without slowing economic growth. They envision a society in which "clean" high-technology industries will provide the material basis for a society in which more and more Americans will live by "processing information." Before wrapping-up this discussion of environmental issues, it is necessary to address the myth of clean high-technology industry.

The myth of "hi-tech" depends upon the availability of a never ending supply of increasingly sophisticated computers and other electronic devices. But electronics manufacture, underneath its clean facade, is a series of dangerous chemical processes. Communities dominated by the computer industry are plagued with severe groundwater contamination from cyanides, arsenic, toxic heavy metals and a wide range of carcinogenic chemical solvents, all essential ingredients in the manufacture of silicon computer chips. People living near electronics factories often report strange allergies and reproductive disorders which are mysteriously "cured" once they move away.

The electronics industry claims a rate of occupational illness two to three times the average for manufacturing industries, with frequent complaints of severe acid burns, chemical-induced disorders of the nervous system, kidneys and liver, menstrual irregularities and frequent miscarriages. The most routine and the most hazardous operations are often moved to countries in Latin America and Southeast Asia, where local water supplies are routinely destroyed and young women workers find their eyesight and reproductive health to be permanently damaged after just two or three years of assembling computer chips. The "information society" does not use any fewer goods; it simply seeks to better hide the consequences of their production.

The Santa Clara Valley of California, once a pastoral landscape of farms and orchards, has become Silicon Valley, one of the most congested, smog-ridden areas of the state. Still, local officials from across the United States flock there to view the "economic miracle" of hi-tech. Many of them return home and push through programs of economic incentives and relaxed environmental regulation to promote high-technology development. The consequences are not revealed until their own neighbors begin to suffer the effects of the industry's often careless practices.

At the forefront of high-technology today is the newer field of biotechnology. Biotechnology involves the genetic and biochemical engineering of natural processes on the microscopic subcellular level. With promises of cures for crippling genetic diseases, of self-fertilizing food crops and of human-engineered microorganisms that will perform all kinds of industrial miracles, the biotechnology industry has attempted to evade public scrutiny and public regulation.

Efforts to monitor the uses of biotechnology have originated largely in communities facing the unknown consequences of releases of genetically-altered bacteria into the environment. On two occasions, local fears about the hazards of genetic engineering have sparked debates at the national level. The first was in Cambridge, Massachusetts, in 1976 when Harvard University and the Massachusetts Institute of Technology sought to build two of the first laboratories to be specially-designed for potentially hazardous genetic experiments. Local residents, with the help of some skeptical scientists, were able to place constraints on just what kinds of experiments would be allowed there.

More recently, in Monterey, California, Green activists were at the forefront of efforts to alert their neighbors about the possible dangers of open field testing of genetically-modified soil bacteria. One company in California has created a pair of new bacterial strains that are claimed to help postpone early-and late-season frost damage to a variety of food crops. Little research has been done on the possible effects of these bacteria on local weather patterns, the formation of clouds and the wintering cycles of other plants. In Monterey and Berkeley, Greens have helped generate enough political and legal pressure to prevent field tests from occurring and have been helping to spread the word to other communities that may be affected in the future. Thanks to these local efforts, the broader national debate over genetic engineering has begun to heat up again after nearly a decade of relative silence.

Examples like these illustrate the difficulties of trusting in new technologies as an instrument of positive social change. Technologies are only as good as the society that creates them; the more powerful the technology, the more it can amplify the qualities of the society it was designed to serve. Ecologically-sound technologies are most likely to be created in a social setting that places the highest value on healing the relationship between people and the rest of nature. In such a setting, technology can help to nourish that relationship, aiding in the shift toward local production for local needs, an earth-sustaining bioregional agriculture and a way of life that values conservation over consumption. An ecological society would affirm the idea of "appropriate technology," technology inspired by a life-affirming sensibility, modest in scale and oriented toward ethically-appropriate ends. A society that seeks to manipulate and control nature, on the other hand, will find itself pushed ever-closer to the edge of ecological collapse as the earth's life-sustaining qualities are further and further degraded and social structures are distorted to adapt to new technologies. The German social philosopher Max Horkheimer saw this pattern emerging more than a generation ago when he warned:

> The more devices we invent for dominating nature, the more we must serve them if we are to survive.
> —from *The Revolt of Nature*

Ecological issues are often approached in a rather piecemeal fashion, as environmentalists tend to champion their own pet issue in relative isolation from all of the others. Groups formed around wilderness preservation and endangered species have historically been slow to address politically sensitive issues such as nuclear power and toxic dumping. Public health activists concerned about air pollution and poisoned water supplies have tended to see rural issues as outside their domain. A Green perspective encourages people to uncover the underlying causes of environmental problems in the habits and assumptions of the societies that have created them. At the same time, Green approaches to social issues need to be thoroughly informed by an ecological sensibility. How we do this should become clearer as we explore the problems of social justice and of political and economic democracy in the following chapters.

4.
Social Justice and Responsibility

One of the most persistent myths of industrial cultures is the notion that economic growth can help relieve poverty and inequality. This belief is a cornerstone of both capitalist and modern socialist ideas of progress and it forms the basis of both Western and Eastern attitudes toward the world's poor. The more people become engaged in a cash economy, the more they tend to accept the idea that manufactured wealth, whether in private or state hands, can "trickle down" to improve their own lot.

This trickle-down theory, whether in the guise of social welfare, "supply-side" economics, international development aid or socialist reconstruction, has been a dismal failure for most of the world's peoples. It provides a convenient cover for the dominance of transnational corporate interests and international power blocs in world affairs. Rather than improve the ability of people to feed and house themselves, ward off disease and strengthen their own cultural integrity, the industrial systems have bound people to an entangling web of dependencies totally outside their own control. Sometimes a local elite is created which is permitted to share in the material benefits of "development," but the vast majority of people in "developing" countries are left worse off than before.

Consider the growing problem of hunger, certainly the most profound social tragedy the world faces today. In 1950, Africa as a whole was considered to be self-sufficient in food; today tens of millions of tons of basic cereal grains must be imported from the "developed" world every year. Even under colonial rule, with all its obvious injustices, most people in remote areas had far less contact with the outside world than they do now and were able to maintain their own subsistence economies. Today many countries in Africa, Asia and Latin America produce a greater volume of foodstuffs for export than their people could possibly consume, yet people are starving.

Part of the problem is ecological. Environmental and climatic changes resulting from industrial pollution, the abuse of water resources and the destruction of rain forests are stretching the boundaries of the world's deserts. The borders of the Sahara, for example, are expanding at rates approaching a hundred miles per year.

A far greater proportion of the world's hungry people, though, owe their plight to the same international sources that offer them aid. In country after country, the best agricultural lands have been appropriated for export crops like coffee, sugar, tobacco and tropical fruits, disrupting village-based ways of life that have sustained people for countless generations. People are forced to grow their food on increasingly marginal lands and are obliged to pay high rents and taxes on lands that once belonged to their own villages. Shipments of food aid are increasingly used as a diplomatic weapon to tie impoverished countries to American policies and American markets. Neo-colonial economic relationships force Third World governments to become instruments for exploiting their own people.

In other cases, farmland is permanently disrupted by road construction and new industrial developments. People are compelled to abandon their traditional ways and become factory workers and consumers of goods. The disruption of local economies and local cultures disturbs the balance of family life, accelerating population growth and leaving increasingly desperate masses of people no recourse but to flock to already overcrowded urban centers. If present trends continue, most of the world's people will be living in cities by the year 2000. Whether the change comes by force or through more subtle economic pressures rarely matters in the final analysis; subsistence economies and cultures often collapse within a generation or two. The ecological havoc wreaked by modern technology-intensive developments greatly accelerates the process of cultural decay.

Similar forces are at work in much of rural America, except that the skills and community stability necessary to maintain a local economy were often lost a generation or two earlier. New large factories are increasingly sited in poverty-stricken rural areas where people will work for low wages and labor unions have not established themselves. Such projects are encouraged locally in the name of "economic revitalization" and other lofty phrases, but the boom-town atmosphere that accompanies such developments brings in outside workers, strains local services and rapidly drives up living costs. What has survived of a local culture is often shattered, driving local people deeper into poverty and dependency. At the same time that American farmers are being paid not to grow corn and to kill off their dairy herds, hunger is rampant in our inner cities and in growing numbers of rural areas.

Greens view basic human needs such as food, shelter and health care as fundamental social rights to which everyone is entitled. These are the basic survival needs once guaranteed to members of tribal societies, regardless of a person's age, physical strength or social status. To this list of rights we would now add guarantees of basic political freedoms and an overall sense of social stability and support from infancy through old age. The idea of people having to compete in the marketplace for the privilege of eating is an archaic holdover from the disruptive early phases of European capitalism. It has no place in a society that pretends to value the sanctity of human life.

The basic instrument of social support is the community, whether that be a city or a village, an urban neighborhood or a rural town. Decentralized, face-to-face institutions are in a better position to understand people's needs and assure that they are satisfied than the faceless public welfare bureaucracies that governments worldwide have created. The current welfare system makes people slaves to a humiliating bureaucratic machine that systematically destroys its clients' sense of self-worth and personal initiative. A network of decentralized self-help programs on the community level can take much more meaningful steps to reverse the slide into poverty and despair.

We are speaking, of course, of real decentralism, founded upon bioregional self-reliance and genuine community control. The so-called "federalism" of the early Reagan years, with endless rhetorical and budgetary attacks on the federal social service bureaucracy, was merely a coverup for a massive diversion of public funds to the military and to corporate coffers. Rather than empowering communities to take on the support roles being whittled away from the large federal agencies, the resulting policies have increasingly deprived states and localities of the resources necessary to provide basic services. Fully two-thirds of our federal tax dollars are being spent on military and military-related programs; the largest share in history of those tax dollars are being paid out by individual taxpayers as opposed to corporations and businesses. The much-touted 1986 "tax reform" makes little difference. Even the meager revenue sharing with cities and states initiated during the Nixon years has been swallowed up by the billion-dollar-a-day military budget. Reversing these damaging trends needs to be a first priority. Whatever resources people contribute to help maintain the social infrastructure need to be allocated in a locally responsible and democratic manner.

Re-equipped with the resources currently being drained from them, communities and federations of communities could begin to cure poverty and social decay. The wealth of existing self-help efforts, from emergency food shelves to volunteer fire departments, from housing rehabilitation projects to hospice programs to neighborhood crime watches, offer models for a truly face-to-face, neighbor-to-neighbor approach to social problems. The outcome of these combined efforts would be the creation of a truly social safety net, grounded in a spirit of human solidarity and mutual aid.

Many experiments in different parts of the country have shown how decentralizing and de-institutionalizing basic social services can provide better care for people at much lower cost. Innovative projects in early childhood education, health maintenance and care for physically and mentally handicapped people have shown that people's needs can be far better served in more intimate, personalized settings that simultaneously encourage greater ties to the community at-large.

An ecologically-minded social outlook would also strive to relieve the special oppression felt by many groups of people in our society based upon their race, sex, age, nationality or sexual orientation. An ecological society would embrace the diversity and unique contributions of all the many types of people that share the earth together. Embracing social and cultural diversity is an important step away from the provincialism and isolation that often plagued village-based societies in the past.

It has never been possible to combat racism and other forms of prejudice by pretending to ignore the differences that exist among people. Our society's attempt to homogenize everybody through a sieve of mass culture simply reinforces feelings of alienation and isolation. The fear of having to compete for one's share of material resources against those who appear different drives people to mistrust their closest neighbors. A society informed by Green values would enhance the opportunities for people to work together to satisfy their real needs, as opposed to the limitless but unattainable desires created by consumer society.

In West Germany, the Greens have taken up the cause of migrant workers from Africa and the Middle East, who have come under increasing attack from the far right. The racism and national chauvinism being stirred up in these efforts to isolate people of color has aroused bitter memories of the Nazi era. The economic program of the New Haven, Connecticut Greens (see Chapter 7) makes a priority of supporting the needs of the black community there. A   social outlook embracing diversity calls upon people to confront both the institutions that support racism and the cultural insecurities that give these institutions their legitimacy.

Perhaps the most pivotal form of oppression in our society is the continued domination of women by male-dominated social, economic and even familial institutions. From our tribal past, in which the male and female social spheres were sometimes separate but always complementary, our civilization has built a patriarchal culture that still systematically devalues women's contributions and women's selfhood. The domination of women goes back at least as far as the earliest warrior societies and has long transcended merely economic forms of oppression. With the increasing denial of people's ties to nature in hierarchical societies, women became a symbol for what is fearsome and uncontrollable in the world. The suppression of nature and the domination of women have become two sides of a single life-denying cultural impulse.

As women's oppression most clearly goes beyond the economic and political sphere, the feminist movement has evolved the clearest understanding of the deeply personal side of social inequality. Many feminists have called for a reexamination of the whole idea of power, both in society and between individuals. Ecofeminism (see Chapter 2) represents probably the deepest current expression of a personal tie to the natural world and the need to reflect ecological principles in even the most intimate aspects of daily life.

Can a Green social vision speak to women's need to break through millenia of oppression and fully realize a renewed sense of selfhood? The first step is for Greens to fully support ongoing efforts of women to assert their place in our present society. This includes the right to equal pay for equal or comparable work, equal rights under the existing legal system and the right to choose whether or not to bear children. Workplaces should be obliged to support high quality day care facilities. The special dilemma of single mothers, now the largest single category of poor people, demands greater attention and community support.

Women's safety on urban streets needs to be guaranteed through neighborhood crime watches, citizen patrols, networks of safe houses and other community-centered strategies. The emotional support work that often falls upon women in organizations, workplaces and at home needs to be fully valued and shared by everyone. All of the cultural institutions that reinforce women's subordinate status need to be confronted, devalued and replaced with life-affirming alternatives.

Each of these steps is crucially important, but institutional changes are only a beginning. Attempts to better integrate women or other oppressed groups into a competitive society are at best merely holding actions on the road to more thoroughgoing changes. The values of nurturance, cooperation and sharing which are traditionally identified more closely with women than with men need to become the deepest underlying principles of our society. Many women clearly need to be freed to become more assertive and self-confident, but the embracing of competitive, power-seeking roles by women aspiring to "make it" in the corporate world largely reinforces the strength of patriarchal institutions. Successful corporate women become the exceptions that prove the rule; their images are especially useful in the effort to glamorize corporate culture.

The bioregionalist emphasis on home and community offers one possible focus for rethinking traditional role stereotypes. Canadian feminist Judith Plant has suggested that bioregionalists seek to re-value the domestic sphere as the focus from which men and women together raise children, grow food and participate in public deci-sions. The home—and the extended family, whether biological or voluntary—is thus transformed from a place where women are trapped to a focal place from which actions in the larger society originate. Says Plant,

> We are attempting to move out of culturally-defined sex roles which value one over the other, toward a culture that places posi-tive value on the active involvement of all people in domestic life. For it is here where culture is shaped.
> —from *The New Catalyst,* Number 2

Work and Technology

Greens worldwide have called for a restructuring of work and a new social relationship to technology. Earlier in this century, the abolition of the wage system, and its replacement by a new industrial order in which workers would fully control their own labor, was the stated goal of many sectors of the labor movement. This vision was abandoned in the post-World War II years by a labor movement that was increasingly secure materially and increasingly accommodated to the existing system. Most working people came to accept the idea that incremental increases in wages and benefits were the only attainable ends.

The economic uncertainties of our time, the dramatic rise in the accepted level of unemployment and the rise of new technologies in the workplace are again raising questions about the nature of work. Workers in basic industries in the United States have agreed to cuts in pay in the hope of keeping their jobs. Engineers are devising elaborate schemes to replace manufacturing workers by robots—as many as possible within the next generation. The jobs that remain, both blue collar and white, often become increasingly monotonous as they are tied more closely to the pace of machines.

The myth of computers freeing people from toil has proven to be far from the truth. Instead of lightening the burden of work and freeing people for more creative pursuits, computers and robots have made life both on and off the job increasingly insecure. Technologies that could be enhancing creativity are more often suppressing it. As machines appropriate many of the manual and intellectual skills that once gave workers their sense of pride, decisions about how tasks are performed are increasingly reserved for the programmers and managers. Workers' skills become obsolete at a faster rate and training in new skills is rarely available. Skilled industrial workers often find themselves channelled into low-paying unskilled jobs in the service sector.

Office workers confined to computer terminals for their entire work day are often monitored for speed and efficiency with every keystroke. Engineers and architects working for large firms are often made to feel as if even they have been put on the assembly line. When people protest the increasing alienation and lack of control they feel on the job, managers respond with artificial gimmicks such as "quality circles," borrowed from the Japanese. Such methods often help managers design a more efficient production scheme, but they are rarely a forum for workers' genuine grievances.

Work does not have to be boring. Work can be the highest form of creative expression, a real vehicle for everyone to contribute their skills for the betterment of themselves and their own community. People rarely shirk jobs that offer a real sense of personal fulfill-ment. Where the sense of community is strong, people seldom hesitate to commit their precious free time to projects designed to make everyone's life better. How often do we find the work we do for money interfering with the more fulfilling and productive work we wish to do on our own?

A society committed to satisfying basic needs should not compel people to work. A system that encourages local production for local needs should allow people to pursue a balance of productive and creative pursuits, with both fully valued by the society at large. People's ability to thrive and be creative should not depend upon how much work they do, nor what types. As in pre-industrial socie-ties, part of a week's work should be quite sufficient to satisfy one's survival needs, leaving ample time for other pursuits. The rigid division between work and community service can dissolve rather rapidly as work is reharmonized with community life and the control and ownership of productive resources come to be shared by workers and their communities.

The problems of chronic unemployment and industrial automation are encouraging many people to consider real alternatives to the present wage system. Even people with close ties to the labor movement are questioning the perennial call for government policies to create full employment. Many, including the Greens in West Germany, have proposed an immediate shortening of the work week and a curtailment of overtime to spread available jobs among more people. A socially-guaranteed income for everyone would immediately free people to do some of the necessary work that is not sufficiently valued by the corporate economy—the work of rebuilding communities, restoring the ecology, planting gardens, creating art, and other steps toward enhancing the quality of life for all.

The German Greens have emphasized the need for people to be able to decide "what is produced, how and where"; all economic decisions need to be made by those most affected by them. Their program decries the division of labor and the over-specialization imposed by the wage system. In addition to offering training for all workers to the highest levels of skill, they propose that manual work always be rotated with time spent in planning and supervision of the productive processes. Freely-available educational leave, shorter working hours and the protection of labor's right to organize are also important short-term goals. When a corporation chooses to close down a facility that is producing goods valued by the community, the workers should be aided in buying out the plant if they wish, restructuring its operations and continuing to operate it under their own management. If the products are not socially useful, or the work environment is especially oppressive, resources should be available to shut it down and help workers rechannel their skills elsewhere.

The West German program also proposes several measures to begin regulating technology in the workplace. New industrial technologies should be evaluated by everyone who might be affected by them, both within the workplace and in surrounding communities. Changes that might increase the level of workplace stress, either physical or mental, should always be avoided. The use of toxic industrial chemicals should be rapidly phased-out. New technologies or uses of automation that reduce people's control over their own working situations should not be implemented, no matter how much more efficient they may appear to be from a managerial perspective.

In the workplace, as in society at large, Greens are questioning the whole idea of technological progress. As in earlier times, those who question technology are often dismissed as "Luddites," after the mill workers of the first Industrial Revolution who roamed the English countryside smashing the machines they felt were threatening their way of life. These people—who marched under the banner of a mythical Ned Ludd—were reacting against the social dislocations and the loss of personal autonomy imposed by the new factory system. In a time when the social changes being heralded in by new technologies might equal those of that earlier era, it may be necessary to ask once again, what purposes do these new technologies serve and why do we need them? If a new machine serves no meaningful function except to further consolidate management control over the pace of work and the production process, why should people be expected to cooperate with its implementation?

It is necessary to defuse the myth that such developments are historically inevitable. In Europe, workers in industries being rapidly transformed by more intensive automation—metalworkers, telephone operators, government office workers, printers and dock workers—have actively resisted the installation of unwanted technologies and continue to back up their opposition with a renewal of direct action on the job itself. In West Germany, Holland, Italy and the Scandinavian countries, unions have fought for veto power over new workplace technologies, and there have been a few important local victories, especially for white collar workers. Where technicians and engineers are organized into unions, they have been known to contribute their skills to render offending technologies unusable. Some industries in the United States, particularly the automobile and aircraft industries, face similar opposition to the imposition of disempowering workplace technologies.

For historian David Noble, such active resistance in the workplace is in the best tradition of the original Luddites. It is a demonstration of "the supremacy of society over mere economic activity and technological contrivance." In the 1940's and '50's, many of the forebears of cybernetic technology urged that the pace of technological change be slowed in deference to social needs. Now, even professional computer designers are beginning to fear that their own creations will take their jobs away. As Rudolph Bahro once said, "The problem is not the abolition of technology, but its subordination" to social needs.

Housing, Health, Education and Culture

The economic changes of the 1980's have made it more difficult than ever for Americans to find adequate housing. From the 1950's through the seventies, government loan guarantees, interest subsidies and tax breaks encouraged housing construction. The construction industry is still sustained by federal subsidies, but now only commercial developers are eligible for most of these advantages. The result is a declining housing stock and a marked increase in real estate speculation. Housing costs in many cities have doubled and tripled within the last decade. People are often driven out of homes they have occupied for many years by upscale professionals seeking high-consumption urban lifestyles, a process known as gentrification. The Institute for Community Economics in Greenfield, Massachusetts, has reported that proportionately fewer Americans can now afford to buy a home than was the case during the Great Depression.

One Green approach to housing centers on the idea of the Community Land Trust. Community Land Trusts are non-profit organizations which acquire land and buildings in the name of the local community. They are funded by donations or public funds, but they can quickly become financially independent through their holdings. Houses and apartments are leased or sometimes sold to their tenants, but the Trust itself retains the rights to the land, thus removing it forever from the speculative real estate market. Such efforts can be initiated by groups of individuals, neighborhood associations, charitable organizations or by cities and towns as a whole, and are designed to encourage widespread participation by people in a town or neighborhood. Once the land itself is removed from the pressures of speculation, people can feel more secure about the future of their own home, whether privately or cooperatively owned. Land Trusts can provide the personal security that comes with home ownership, with the value of the land and of resources offered by the community retained by the community as a whole.

Starting in New England, many localities are also establishing Community Loan Funds to enable low income people to own their own homes. These revolving funds also can be initiated by private or public institutions; churches often provide initial seed money. People's community ties are more important than their income level in deciding whether they qualify for a loan. Loans are offered at interest rates well below those available from banks, and every dollar paid back on an existing loan becomes available for another family or cooperative seeking to establish themselves.

The concept of sweat equity is another innovation of great importance, especially in inner cities where once-livable houses are often deteriorated or abandoned. Under sweat equity, the value of the work that goes into restoring an old house can accrue as equity toward the eventual ownership of the building by its tenants. In some cases, this has even allowed people who have moved into abandoned houses as squatters to create more permanent homes for themselves; in all cases, it helps to revitalize once-decaying neighborhoods, while building people's skills and self-esteem. In tandem with Land Trust ownership, sweat equity can help long-term neighborhood residents resist the pressures of gentrification.

An ecological approach to housing should also consider the importance of green spaces in the inner city. Many of the first urban Land Trusts were created to acquire vacant lots for the creation of neighborhood parks and community gardens. The opening up of green spaces, especially where food growing and edible land-scaping can be practiced, is an important step toward relieving urban alienation and uniformity.

Countless volumes have been written about the inadequacies of our present health care system. The skyrocketing cost and the deteriorating quality of care, the treatment of symptoms rather than causes, the pushing of drugs and technological solutions have all come under increasing scrutiny in recent years.

Greens understand that a healthy society is necessary for the development of healthy people. The body is a microcosm—the same ecological principles apply to its inner workings that apply to nature as a whole. An ecological approach to medicine looks at every person as a whole being whose state of health is intricately intertwined with the health of the planet.

Greens look at the dramatic rise in chronic diseases such as heart disease and the many forms of cancer as a sign of our society's failure to sustain the earth's ability to support life. The West German Greens' first national program lists some of the specific causes:

> The anti-human methods employed in offices and factories . . .
> The disturbance of ecological balances by air and water pollu-
> tion, radioactivity, denaturing of food, poor nutrition, weakening
> of the body's own healing processes by symptom therapy,
> mental stress, deprivation of meaningful human relationships,
> excessive use of medicines and misuse of drugs.

Adequate health care is a social right, not a privilege to be bought by those who can afford it. A renewed health care system requires widely accessible local clinics that emphasize the prevention of disease. A full range of services, including nutritional counseling, a choice of alternative therapies and the teaching of methods to relieve stress, would be available to help people stay healthy. Disease therapies should seek to restore the body's natural defenses instead of mounting a chemical assault on specific symptoms at the risk of more long-term damage.

More sophisticated methods of diagnosis and treatment could be available at regional hospitals in those cases where locally-available methods are not adequate. Some form of public insurance will be necessary to guarantee the availability of such facilities, though a National Health Insurance system as often proposed would likely reinforce the present trend toward the centralization and increasing technological intensiveness of hospital care. A sensible approach would first eliminate the commercial pressures on hospitals to acquire prestige by competing for new technologies. Technological overkill and duplication of facilities are now major factors in rising health care costs. The control of the large drug companies over research, publishing and education must be eliminated, and medical education itself needs to be restructured to evolve a more holistic-minded, service-oriented corps of health care providers.

Our hopes for creating an ecological society rest to a great extent on our ability to change the way we educate our children. Standardized compulsory education is the product of a past era when social reformers sought to mold large numbers of new immigrants to the ways of urban American life. This process continues today, as schools continue to encourage conformity, subservience and competition. Rather than helping students realize their potential to become creative members of their community, most public schools are highly regimented and confining. They encourage young people to recite correct answers rather than ask new questions. In rural areas, the homogenization of local community schools into large unified school districts has merely reproduced many of the chronic problems of urban schools.

Greens advocate diversity and variety in education. Children should be allowed to break free of stifling institutionalized schools and immerse themselves in their community and in nature. Social support should shift from heavily bureaucratized central school districts to community groups and parent/teacher-centered cooperatives.

Rather than having to sustain large buildings to lock young people up for six to eight hours a day, classroom instruction in basic skills should be combined with time spent out in the community, learning and working with local craftspeople, farmers, artists, scholars and professionals. Even young children have been shown to learn faster and better when they have real choices about what to learn and how. Any gaps in what is available for students in a particular community could be filled by a free and open program of exchanges between nearby communities. It is especially important to encourage educational exchanges between rural and urban communities, to cultivate a fuller appreciation of both the ways of nature and the diversity of human cultures. Wilderness education and experiences should be available to everyone at an early age.

Many religious groups and individuals identified with the political right have been pressing for an educational voucher system to replace standardized schooling. Under such a system, parents who wish to remove their children from public schools would receive government vouchers redeemable for tuition at other types of schools. This idea has been criticized as a step toward lowering educational standards and toward government subsidy of religious institutions. Clearly, that is what many current advocates of the voucher system would like to see, especially the fundamentalist churches.

What if we could come up with a plan that gave no particular advantage to large established institutions such as religious schools, but instead encouraged a wide diversity of experiments in alternative education? The most authoritarian elements could be discouraged in a number of ways, for example by requiring that schools be cooperatively run by parents, teachers and students in order to qualify for full public support. Very close scrutiny would also be necessary in the near-term to resist the excessive influence of corporations, the military and other such institutions on the education of young people.

Changes in education, where they succeed, could be the forebears of real changes in the way people come to view their role in society. American culture has evolved in a direction that encourages obedience, subservience to authority and an all-encompassing consumerism. We are raised to be highly competitive at a very early age and to view our personal needs in terms of the most narrowly defined self-interest. Such a system allows some people to thrive, especially those whose social background puts material resources at their disposal. For many more people, however, early childhood experiences shatter their sense of personal confidence and playfulness. Children are taught to forget the pleasures of cooperation and suppress the urge to control their own destiny.

Not only the educational system is to blame. The mass media and all of the cultural signals we internalize while growing up help to create feelings of distress and inadequacy. We are taught to disguise these inadequacies by oppressing others and by judging ourselves by what we own and what we consume.

Television is clearly the source of many of these mass cultural manipulations. It is hypnotizing people everywhere into one homogenized consumer culture. It is intrinsically anti-social, it brings an inflated system of manufactured needs into nearly everyone's home and it offers only the emptiest of role models. No matter what the specific content of television programming may be, protracted viewing breeds restlessness, impatience and a distorted time sense that creates a chronic sense of boredom.

The widespread use of microcomputers in schools and in the home threatens to compound the destructive social effects of television. Computers can make some tasks, such as typing and editing this book, much easier and more pleasant. However, the acceptance of computers as a cultural panacea threatens to raise our isolation from the natural world to new heights of abstraction. School systems all across the country have, in recent years, abandoned more experiential programs in the arts and outdoors to jump on the computer bandwagon. A myth of "computer literacy" has been created by computer enthusiasts to compel parents and teachers to offer computer instruction to very young children.

The real impact of early computer use on many children is a reduction of experience to what can be represented on the screen and an invitation to view the world as an object for manipulation and control. Children pushed to demonstrate formal computer skills at an early age may be impaired in their emotional development. Though some educational computer programs are now designed to encourage more originality, most school systems are embracing "computer literacy" as yet another regimented "basic skill" that reinforces the stratification of students. In school and throughout society, computers have become a vehicle for compounding the drive for efficiency, reinforcing social regimentation and further devaluing all aspects of our experience that cannot, and should not, be expressed with numbers.

The myth of the "information society" promises a vast increase in the availability of information and an enhanced freedom of choice. Instead, the more information becomes a commodity to be bought and sold, the more it can be controlled and channelled by those who have the money to buy it. The more corporations create their own computerized libraries of information, the more pressure we see to cut funding for public libraries. Both corporations and the government are also greatly aided by computers in their efforts to monitor our behavior and our patterns of consumption. In society at large, as in the classroom, we face the prospect of a deepening inequality among people based on their ability to purchase and use computerized information.

Greens understand that much of what we seek to change in society is deeply ingrained into our culture and our language. Living in a highly competitive setting where greed and dishonesty seem to pay makes one feel as though competition is at the core of human nature. But think for a moment about all of the people you admire for their willingness, when necessary, to put their own needs aside to help others. Think about all of the things that make life in your town special that depend upon people's simple generosity and sense of giving. When people are able to work together closely to brighten up their lives and those of their neighbors, cooperation, not competition, is usually the rule.

We need to free ourselves from the institutions of mass culture that reinforce competitiveness and people's feelings of personal inadequacy. We need to encourage a flowering of local culture and local celebrations. People will only unglue themselves from the television screen when they can find activities and creative outlets in their own community that are truly fulfilling, that foster a closeness to one's fellow humans and to the earth, that satisfy the genuine need for love and acceptance so often sublimated in the pursuit of manipulated artificial needs. There is no formula for social and cultural change; trying to guess which comes first is like the proverbial problem of the chicken and the egg. What ecology, and a Green outlook, have to offer is an affirmation of everyone's uniqueness and the understanding that each of us is strengthened by the depth of our shared interdependence with our fellow beings.

5.
Democracy in Politics and in the Economy

Greens both in Europe and in North America are questioning all the imbalances of power that plague the world today. A genuine commitment to democracy requires that we seek alternatives to both the capitalist and state socialist systems of centralized power.

In the West, our profit-oriented system allows a privileged few to control the destinies of millions of people by accumulating control of capital, the productive resources that make our economy possible. Intense, open rivalries over how to best manage the system abound, but every opportunity to increase the concentration of power is fully embraced by those who control the wealth, no matter who sits in the White House.

In the state socialist countries to the East, power is even more heavily concentrated. A massive self-perpetuating Party bureaucracy rules, with different agencies and political sectors competing for shares of power and resources. Internal rivalries are more covert, but more explicitly political; the goal again remains the maintenance of the system.

All sorts of variations on these basic themes exist in the industrialized world. In the "underdeveloped" world, more overtly repressive regimes are placed in control in order to keep the resources flowing into the hands of the superpowers. No matter how democratic people's goals and intentions might appear, the centralized state and centralized world economy inevitably concentrate political and economic power in very few hands.

A decentralist alternative needs to evolve, in which people directly make the decisions that affect their lives. Politically

97

independent, bioregional communities can offer a setting in which people can carry on their affairs in an open face-to-face manner, offering everyone a full say in decisions that affect their lives. Such communities could form the underlying basis for an ecological society.*

A political transition to face-to-face local democracy and an economic shift toward local production for local needs are two sides of a Green strategy for changing society from the grass roots. A third component would be the joining together of free communities to exchange goods and solve common problems. The interdependence of communities can help enrich the lives of their inhabitants, but only if each community is an autonomous, self-sustaining whole. Federations of communities could be entirely voluntary, formed for specific purposes and subject to direct popular control. In time, they could become a countervailing power against that of the centralized state and powerful corporations. Such alliances are also necessary for correcting imbalances of material wealth that might otherwise pit communities against one another.

Probably the best-known institution of face-to-face popular democracy in our time is the traditional New England Town Meeting. For two hundred years, people in New England towns have gathered together toward the close of winter to discuss issues of local importance. The decision-making powers of Town Meetings have dwindled over the years as state and federal agencies have captured jurisdiction over many local concerns. However, at least in the more northerly rural areas, towns are still sovereign

---

*In a society as mobile as ours, recreating the necessary sense of community might be the greatest stumbling block. That social mobility, however, allows for a depth of voluntary associations among people that helps break through the isolation and provincialism of many traditional communities. Bioregionalism and the goal of community self reliance both offer ways to reclaim a geographically-based sese of community, but it is not the only way. In today's world, the community one counts on for support might be united more by common interests or experiences than by geography. A few writers have asserted that the "nomad" or the uprooted "masterless" person may have played a crucial role in the evolution of human freedom.

when it comes to managing roads, schools and firefighting equipment, setting property tax rates and deciding how to spend those funds, supporting local social service agencies and other functions. In recent years, Town Meeting discussions of broader issues such as the nuclear freeze, acid rain and nuclear waste have helped set the stage for regional and national debates.

Town Meetings are by no means a perfect institution. Parochial interests and the entrenched power of long-time landowners can stifle full participation. Annual attendance has dropped over the years as local control is increasingly compromised by directives from other places. However, Town Meetings, and the active, face-to-face political culture they symbolize, represent a starting point for a decentralist politics grounded in genuine local control. Even when power is usurped or when the big picture is lost in a swarm of details, people know that what they say matters. If local officials elected at an open assembly make an unpopular decision, their neighbors can, at the very least, guarantee a few sleepless nights.

Participatory political forms have also been successful in urban settings. Governance by open public assemblies was the cornerstone of ancient Greek democracy, the Parisian sections of the French Revolution, pre-Revolutionary Boston and the anarchist city of Barcelona during the Spanish Civil War. Greens are just beginning to appreciate the importance of these past urban experiments to our hopes for evolving a new politics.

Beginning in the 1960's, many cities began establishing neighborhood-based forums for citizen participation, especially around long-term planning issues. These bodies are generally restricted to an advisory role and are often dominated by aspiring politicians. In many cases, though, they are a possible starting point toward real neighborhood control. The size of the community and its population density are not as important as the sense of human scale that allows people to work together closely as neighbors.

In Burlington, Vermont, a group of radical ecologists pressed for the creation of Neighborhood Assemblies as a step toward democratizing the city government. The mayor, a Socialist with broad popular appeal, reluctantly embraced the idea when his re-election strategy required more evidence of public participation in city government. The original proponents' full vision—to transform the City Council into a coordinating body of delegates from the Assemblies—still remains a distant one. However, the Assemblies have had significant influence around local development issues, the apportionment of federal Community Development funds and the voicing of neighborhood issues and concerns.*

In an urban setting, especially in smaller, more human-scaled cities, the popular assembly form has the potential to evolve to an even higher level of participation than exists in rural New England. Rural people are often reluctant to discuss any but the most immediate public issues, such as roads and school budgets, at Town Meetings. Discussions of other matters, even those with significant local impacts, such as land use and development, are often seen as intrusions on people's private lives. This is a marked change from the early colonial era, when the allocation of plots of land to different purposes was often decided communally.

Reluctance to engage in the public sphere partially reflects rural people's sense of personal self-sufficiency and a healthy desire to prevent life from being swallowed up by politics. On the other hand, maintaining a high quality of life in times such as these can require real foresight, given the ever-mounting pressures of land speculation, over-development and environmental degradation. The old American devotion to private property often leaves rural communities more open to manipulations by powerful outside interests that can threaten a community's very survival.

In more populated areas, from small villages to large cities, people are more dependent upon public works to support their domestic needs. One steps into a public sphere as soon as one leaves the home. However, in an urban setting, the spirit of self-reliance is generally missing. Here, people tend to leave decisions more in the hands of politicians and professional experts, from plumbers to city planners. An ecological model of community self-governance would combine the best of rural self-reliance and urban interdependence in a system of real popular control.

Greens are acutely aware that no community is an island unto itself. There is a big difference between self-reliance and isolationism. Community self-reliance requires more than a sense of personal self-sufficiency. Communities need to evolve new ways of

---

*For more details on the unique situation in Burlington, see Chapter 7.

federating to meet common needs and share culture without reproducing the top-down authority structures that now plague us. As soon as a bureaucracy is created, it begins to evolve its own spheres of power and influence that encourage dependency and further centralization.

In an ecological model, the initiative for people and communities to join together would always originate at the local level. Alliances formed for specific purposes are free to form and disband, as local needs change and develop. Before an important decision is binding on member communities, people should have the right to affirm or deny the decision at the local level, with everyone affected by the decision guaranteed the right to be heard. Some people will always choose to be more active in particular community affairs than others, but the strength of participatory democracy comes with each person's awareness that their voice will be heard whenever they do choose to speak.

Along the short stretch of seacoast that forms the eastern edge of New Hampshire, Greens have been working with their local town officials to form a Seacoast League of Cities and Towns. The issue that brings people together is the anticipated opening of the Seabrook nuclear power plant, possibly the most controversial nuclear power project in the United States. In the winter of 1986, many of the towns within ten miles of Seabrook joined in refusing to participate in a federally-sponsored evacuation drill, a drill required by the Nuclear Regulatory Commission before a nuclear plant can be licensed to operate. People across the border in Massachusetts have pressured their governor to reject evacuation plans for their towns, a decision that has helped re-establish nuclear power as a politically important issue throughout the two states. Meanwhile, the Seacoast League has established strong working ties between communities that should help encourage other forms of bioregional cooperation.

Some writers with Green sympathies have been advocating a form of democracy in which the use of the mass media takes the place of face-to-face meetings. Instead of bringing decisions to the community level, they would have everyone stay at home, watch public debates on television and cast their votes by typing into a computer console. This notion of cybernetic democracy has the appeal of allowing instant opinion polls and referenda on matters of widespread importance, but it ignores some important realities.

Improved communication is an important part of any strategy for political transformation. However, the mass media, by separating politics from its social context, leave people open to an ever-increasing degree of manipulation and ideological control. In the privacy of one's home, there is no real give and take, no forum for challenging assumptions, just an assault of packaged media images.

The New Right and the Christian fundamentalists have clearly demonstrated how careful manipulation of mass cultural images can channel even grass-roots power to serve repressive, life-denying ends. People's democratic impulses and deep-felt craving for community are subverted by the political evangelists—and by all varieties of mass-media manipulators—in ways that ultimately strengthen the instruments of cultural and political repression. Instead of reinforcing the power of the mass media, Greens can help to create community institutions that are self-consciously removed from the manipulations of mass culture and mass society. Changes in the political structure of society will not by themselves usher in a Green future. However, efforts to return politics to the local level can help relieve the feelings of alienation and powerlessness that keep people divided, and renew the habits of cooperation and community self-governance upon which an ecological future could be based.

Democracy in the Green Movement

One of the founding principles of the Greens in West Germany was that the internal structure of their organization should offer a living example of the kind of democracy they sought for society as a whole. They worked hard to make sure elected delegates would be directly accountable to local Green organizations. State chapters are autonomous, with national policy decisions ratified at large assemblies of delegates. Positions of responsibility are rotated frequently, and people holding "leadership" positions in the elected legisative bodies are temporarily barred from positions of responsibility within the party.

In the United States, the decentralized structure of many regional anti-nuclear and peace groups offers an important model for human scaled confederal structures. Beginning in the 1970's, activists began trying to relieve the feelings of fear and isolation many people felt at large demonstrations by helping people organize into affinity groups. These relatively loose-knit groupings of a dozen or so people would prepare for demonstrations together, learn nonviolent methods and consensual decision-making and initiate projects of their own to help educate their neighbors and dramatize their concerns.*

By the height of the anti-nuclear power movement in the late 1970's, autonomous affinity groups had begun to shape the underlying structure of the movement itself. Instead of decisions being made by central steering committees of core activists, many activities were initiated by affinity groups and coordinated on a regional basis by rotating spokespeople. Flaws emerged in this process and changes were made along the way, but many large demonstrations involving thousands of people, major educational and lobbying efforts, legal strategies, and a variety of other long-term campaigns have all been planned and carried out in this manner.

As of this writing, over a dozen long-term affinity groups have established themselves in the hills of Northern Vermont. They are working, each in their own way, to help end the war in Central America, close the state's only nuclear power plant, expose local instruments of the military establishment and create more empowering ways of working together. Coordinating activities is sometimes quite difficult, as people live far apart and each group has its own ebbs and flows of activity. But when the affinity groups do come together, their impact is unmistakable, whether on the streets of the state capital, the halls of Congress or the plant gates of

---

*The name affinity group derives from the *grupos de afinidad* that were at the forefront of the broadly popular anarchist movement in pre-Civil War Spain. Through the 1920's and '30's, the anarchist *grupos* actively confronted the institutions of the Spanish monarchy and helped people to organize their communities according to cooperative principles. By the outbreak of Civil War in 1936, people in many corners of the country had severed their ties to the ruling institutions and were cooperatively managing their communities, farms and workplaces.

local military contractors. Similar affinity group networks have developed in California and, to some extent, in other parts of the country, offering hope for a more sustained, multi-issue peace movement, and serving as centers of personal support for resisting the manipulations of mass society.

Most affinity groups and many Green groups, both in Europe and in North America, operate according to a consensual decision-making process, in which each member present at a meeting needs to offer their consent before any action can be taken. Decision-making by consensus was originally adapted by peace and anti-nuclear activists from the processes developed by American Quaker councils over the past three centuries. Its present form incorporates ideas from feminism and other personal growth-oriented movements. Consensus decision-making does away with the neglect of minority points of view that often comes with voting. Rather than pressing people to choose sides, people are encouraged to follow a path toward a synthesis of different points of view:

> Everyone has the opportunity—some would say respon-sibility—to say what they think. As each person speaks, every-one's understanding of the situation deepens. The discussion is continually redefined and reworked to assimilate each person's ideas and feelings. The options are not pre-arranged in a fixed pattern and the fluctuating process of reaching a decision includes everyone. This process is crucial and just as important as the decision.
> —from *Greenham Women Everywhere*

In small groups, with a high level of shared personal commit-ment and unity of purpose, a relatively pure form of consensus can be desirable. The group seeks to act as one and elaborate its group identity. Any individual who objects in principle to a direction the group is taking can block the group's decisions; thus the group is compelled to fully accommodate individual concerns. This affirms everyone's personal stake in carrying out the actions of the group and helps protect everyone's personal needs.

In larger, more diverse organizations, many of the valuable fea-tures of consensus decision-making can be preserved. Discussions are more open and free-ranging than in a parliamentary-style process, personal reservations can be aired and incorporated and the desire to achieve a synthesis of ideas is retained. However, a larger group also needs a method for resolving deadlocks when they occur—otherwise underlying philosophical differences can literally prevent a group from functioning. Many organizations have opted to retain the goal of consensus-seeking, but with a large majority (two thirds to 90%) empowered to break serious deadlocks. The spirit of consensus is protected in a number of ways: waiting for the next meeting before a vote can be taken; allocating time for people on opposite sides of an issue to try resolving their differences in private; or requiring a larger majority to proceed with the vote than is needed to pass the final decision, a way of assuring that even people in the minority agree that an irresolvable deadlock exists.

Consensus is often not the most efficient way to operate, but it helps assure every member's role in determining the group's course of action. It can also help a group discover underlying truths that might conflict with majority viewpoints. In many feminist groups, where personal empowerment is seen as an integral part of the group's political work, consensus décision-making is the central feature of a carefully structured group process.

On both sides of the Atlantic, consensual decision-making and affinity group-centered organizations have helped many thousands of people become empowered to assert themselves politically in the name of peace and the protection of the environment. Much of the present movement against United States intervention in Central America is also organized in such a decentralized manner.

In northern New England, where the grass-roots anti-nuclear power movement was thought to have all but disappeared after the late seventies, people rapidly regrouped in the summer of 1985 to oppose the threatened siting of a national high-level nuclear waste dump in Vermont, Maine or New Hampshire. Federal Department of Energy officials faced hostile crowds of hundreds, and even thousands, of people in every town where they held meetings to defend their site-selection methods. Within a few months, the federal government announced its suspension of plans to explore East Coast sites for high-level nuclear waste disposal. The ability of people to respond so rapidly to this new threat to their communities reflects the degree of personal empowerment fostered by Town

Meeting democracy and the decentralized nature of past anti-nuclear efforts.

Another important Green principle is the rotation of responsibilities. Green delegates in the West German national parliament initially agreed to serve only half of the normal four-year term, after which they would be replaced by an elected alternate. In most representative systems, the personal proclivities of the elected representative determine what priorities are voiced and what kinds of deals are struck behind the scenes. The practice of rotation was to assure that Green delegates speak as the voice of the people who elected them, not as individual pdelegates speak as the voice of the people who elected them, not as individual politicians seeking national power and prestige.

After serving for two years, two of the Greens' twenty-seven parliamentary delegates refused to rotate out of office. Petra Kelly and former NATO General Gert Bastian asserted that they had acquired a level of parliamentary skill and effectiveness that would be lost to the movement if they relinquished their seats as planned. At first, most Greens resented Kelly and Bastian's refusal to rotate, as it reflected just the sort of political careerism they had sought to abolish. But by 1985, the Greens had been changed by their electoral successes and had become increasingly concerned about their electoral appeal. Many felt that keeping two nationally recognized figures in office was more important than adhering to the principle of rotation. They chose to settle for other ways of keeping parliamentary delegates accountable, including placing the two people who were to replace Kelly and Bastian at the top of the list of Green candidates for the 1987 national elections.

Creating Economic Democracy

One of the major obstacles to realizing the democratic ideals of the Green movement is the control of giant corporate monopolies over our economy. The largest transnational corporations have financial assets greater than those of most countries in the world. Corporate control over economic and political institutions effectively prevents people from fully asserting local autonomy and setting local priorities.

An international cash economy, an international credit system and, increasingly, an international division of labor all combine to strengthen corporate control. Instead of fully using our time and energies to help sustain our communities and improve our quality of life, the fruits of our labor are squandered on behalf of this system. Excessive corporate profits, a highly-inequitable tax system, ever-widening wage discrepancies between workers and managers, the secret manipulations of credit and the diversion of resources to fuel the military budget are only a few of the devices that the system uses to maintain control.

The system often appears monolithic and invincible, but recent developments reveal just how unstable it can be. We have discussed how the modern credit system tends to demand unrealizable rates of economic growth. Corporations must increasingly look overseas for cheaper labor and larger markets for their products. Meanwhile, more and more countries are unable to repay the huge loans that were offered to make it possible for them to buy into the system. The profits of large conglomerates and holding companies are increasingly the result of financial dealings and manipulations unrelated to anything tangible they produce.

Such a system might someday collapse under its own weight; alternatively, it might continue to grow like a cancer until everything is swallowed up by it. In either case, the process would be devastating to everyone trapped in the system's web of dependencies. Top-down schemes for stripping away corporate power by nationalization rarely challenge thoughtless growth and often encourage further centralization. Old institutions and practices often reappear in an even more intractably bureaucratic form.*

*The large state-owned energy bureaucracies in Europe and Canada are a case in point. They are the outcome of real efforts to weaken corporate control over energy prices and supplies. However, in most cases, they have adopted the managerial style of private energy companies, seeking to maximize sales and growth at all costs. In Britain and France, they have greatly accelerated nuclear power development through higher government subsidies and through the suppression of public scrutiny under the guise of "national security" and Official Secrets.

No simple formula will suddenly transform either the Western or Eastern industrial economies. Small changes at the local level, however, can begin to whittle away at our own system's ability to control people's daily lives. In the United States, local experiments in community-based economics can help reassert local control, encourage cooperation and provide working models of a different way of satisfying our material needs. As pieces of the system do occasionally falter, locally-controlled economic experiments can help provide a safety net for affected communities and point the way toward a brighter future.

Many ideas have evolved over the past several years suggesting how a new community economics can develop. Some of the basic features have already been mentioned in our discussions of agriculture and urban issues. Community Land Trusts are an ideal vehicle for protecting farmland, creating affordable housing and removing land from the speculative market. Land Trusts are usually incorporated as non-profit, community-controlled institutions which permanently own land and lease it to members for specific, cooperatively agreed-upon purposes. Community Loan Funds offer low-interest loans from a revolving fund to meet local needs in housing or to help start worker-controlled efforts to cooperatively produce goods for the community. Several organizations in different parts of the country are devoting themselves to helping communities establish these kinds of structures.

The food co-op movement provides another important model for a community-controlled economics. From their origins in small buying clubs of a few households that banded together in the 1960's to get their food at wholesale prices, co-ops have grown to become a formidable economic force. Working outward from the local level, they have created a national network of cooperative warehouses, food distributors and trucking firms. They have provided steady markets and sometimes offered necessary seed money to help small worker-owned efforts to grow and process organic foods.

In the 1980's, many food coops have faltered as members' strong initial commitment has become diverted to other pursuits and the commercial food industry has become more aggressively competitive. Many co-ops have folded, but others have dug in and adopted new management methods to become more stable and better able to withstand the pressure of new commercial imitators. It has been a difficult process, and many co-ops have been unable to stabilize their operations without sacrificing their commitment to open participation and democracy. This somewhat parallels the course of an earlier wave of development of consumer and producer co-ops in the 1930's and '40's. Many co-ops from that era survived by becoming more and more like commercial food stores. Still, the co-op commitment to quality products, the benefits of member owner-ship and the availability of natural and bulk foods pioneered by co-ops have helped change many people's attitudes toward the food they buy.

Probably the first experiments in cooperative economic organization within a capitalist economy occurred during the 1830's and '40's in France and England. People of that era came to envision a future in which community-controlled consumer co-ops and worker-controlled production co-ops would federate together to restructure society in the image of the Full Cooperative. Many of that era's experiments also faltered under the pressures of commer-cialization and the consolidation of national economies. But in the face of growing ecological and economic crises, the need for new cooperative models for living and working is greater than ever. We can be encouraged by the emergence of so many new kinds of co-ops over the past ten years, not only to provide food, but also in areas such as energy, recycling, printing and publishing, the building trades, and machinery repair. Cooperative craft shops are helping independent craftspeople in many areas to sell their own wares. Over two hundred worker-owned and managed collectives in the San Francisco Bay Area operate food stores, garages, bakeries, carpentry shops, schools and at least one law firm.

The next step may be to provide vehicles for cooperative efforts to more actively support each other and build a stronger sense of community. In a capitalist economy, individual enterprises are under constant pressure to streamline their operations. Coopera-tive principles such as community control, rotation of jobs and participatory decision-making are often eroded away under these pressures, and co-ops often start looking more and more like the hierarchical, profit-seeking operations they were supposed to supplant. An independent, cooperatively-oriented economic sector sustained by local community-controlled financial resources could help provide the necessary cushion against these kinds of pressures. Direct economic ties between co-ops in different regions and even in different parts of the world would then help foster a wider spirit of cooperation between different peoples.

Two important models, one in this country and one in Spain, offer clues as to how such a cooperative sector could begin to evolve and what it might look like. In the Berkshire Mountains of Western Massachusetts, people affiliated with the E.F. Schumacher Society have created an active community loan fund, an expanding Land Trust and are taking steps toward issuing a local currency backed up by firewood. In Mondragon, in the Basque region of northern Spain, a church-supported technical school and a small cooperative stove-making shop have spawned, over thirty-odd years, an entirely worker-managed local economy.

The Berkshire project is centered around a loan fund, the Self-Help Association for a Regional Economy, or SHARE. It operates out of a local commercial bank which allows funds deposited into special SHARE accounts to serve as collateral for loans granted by the Association. Rather than using the usual financial criteria, which tend to exclude small alternative businesses and cottage industries, loans are granted by a panel of community residents on the basis of an applicant's personal ties and commitments within the community.

In 1985, people began working to establish an independent local currency for the Berkshire bioregion. The local currency is protected from inflation and the whims of the outside economy by fixing the "Berkshare" to the value of a cord of firewood ($100 in 1985). Berkshares are also protected against speculation and hoarding, as each note only circulates for one year, after which it can be traded in either for dollars or for new Berkshares. Many local merchants have agreed to accept payment for goods and services in these notes. Having an inflation-protected currency that is honored locally will encourage people to "buy local" and can help encourage economic relationships at the face-to-face level. A more personalized economy should help to foster a new sense of ethics and a greater commitment to local self-reliance and mutual aid.

Producer cooperatives in the United States often operate in a social vacuum that amplifies pressures to conform to the competitive ways of the larger economy. This is most apparent in the dairy industry where the day-to-day business style of the large milk processing co-ops differs little from that of large corporate dairies. In Europe, however, where cooperatives can sometimes trace their heritage back to the days of truly independent village economies, we see pieces of a different picture. In Britain and the Netherlands, there are producer cooperatives that have survived for several generations as co-ops. In France, where there is a long history of cooperative experiments, there have been several successful cooperative takeovers of companies that were failing under corporate ownership.

In Spain, the largest co-op network in the world was founded in the geographic isolation of the Basque country. Today, the industrial and agricultural cooperatives of Mondragon employ 15,000 local members and local cooperative banks handle some 400,000 accounts. When somebody begins to work in Mondragon, they do not sign a conventional wage contract. Instead, they become members of their co-op, with full voting rights in a General Assembly. Decisions about everything from workplace procedures to long-term economic planning are made in an open, democratic and decentralized fashion. One study of the Mondragon system describes it as a place where control is "vested exclusively in those who work and in all who work." When an enterprise gets too large for decisions to be made democratically, it usually divides into two separate entities, each managed cooperatively by their own worker-members. The regional cooperative bank assumes a great deal of the responsibility for getting new co-ops started, providing much of the needed financial, managerial and technical assistance.

Experiments such as these pave the way for the creation of a new kind of economic system grounded in ecological principles. Time and experience will tell what kinds of economic forms are truly liberating and what kinds simply channel people back toward the dominant system. In a decentralized economy, some communities might choose to experiment with alternative financial schemes while others might choose to eliminate money altogether and create a pure barter system for exchanges of goods and services within the community. A number of hybrid systems are being tried in a few places across our continent.

As corporations increasingly focus on their manipulations of financial markets, there may be increasing opportunities for those productive functions that are worth maintaining to be taken over by cooperative, self-managed efforts at the local level. This will require a variety of inventive schemes to free up financial resources for community use and to recapture local resources currently being squandered by the commercial credit system and the military budget. To promote the transition to a community-controlled, ecological economy, Greens have advocated many specific short-term changes in economic policy.

In West Germany, the Greens have published a detailed economic program, called "Purpose in Work—Solidarity in Life," which calls for an economy that is ecological, socially-responsible and democratic, hoping to reverse the destruction left by decades of unchecked growth. Mass industry would be replaced by decentralized, ecologically-sound production geared toward local needs. The economy would be aimed at enhancing people's quality of life, rather than urging the endless consumption of goods. Unemployment would cease to be a problem, as the necessary work to sustain communities would be justly shared among all who wish to work; the more unpleasant tasks would be shared by everyone.

In the near-term, the German Greens propose a restructuring of work and the development of local self-help programs to assist those in need of social support. A job-training tax would be levied on all companies that do not offer their own training programs to help people to learn new skills. Investment of public funds would be rechanneled along ecological and social lines to assist in the development of renewable energy sources, public transportation, sustainable agriculture and cooperatively-managed public housing. Hazardous and socially useless industries (presumably including the manufacture of weapons) would be converted to more meaningful production, and a variety of alternative projects would receive the social support they deserve. Nonrenewable resources, such as air, water and minerals, would be removed from the control of the market system. The tax system would be restructured to raise rates for the highest income groups and to eliminate corporate loopholes.

Small Is Not Enough

Changing the way the economy works will require a combination of short-term and long-term measures. Some Greens in this country take a much more limited view, simply encouraging small businesses and suggesting that the economy can be decentralized where "appropriate." The more open-minded sectors of the corporate economy are often quite willing to oblige in ways that ultimately strengthen centralized control over economic decisions.

In both Europe and the United States, many companies are taking advantage of new communications technologies to physically decentralize their operations. Information is rapidly moved across the country by computer, allowing different phases of production to be dispersed to more rural areas. This encourages smaller, more spread-out production units, but it leaves even fewer people based in one central headquarters with a complete picture of what the company is doing. Workers are less able to organize and everyone is effectively cut-off from central decisions that affect their future. A large plant employing thousands of rural people becomes just another line item on the corporate balance sheet that can be reorganized or even shut down at will. Many corporations in France, the United States and elsewhere have chosen to physically decentralize as an explicit strategy for silencing union activity. As long as the top-down corporate command structure remains intact, physical decentralization can actually serve to tighten centralized control. An ecological society demands decentralization as a matter of principle, not just for expediency or for the convenience of the existing power structure.

Working to create an economy that can reflect ecological principles is a crucially-important part of any Green movement. It is a slow, evolutionary path with many hazards along the way. Many of our experiments will collapse under the pressure of having to compete with powerful, well-financed institutions. Many others will begin to look more and more like their mainstream counterparts in an effort to appear more "businesslike," until they are thoroughly swallowed up by the commercial sector. However, with every new attempt to free ourselves from the competitive and exploitative ways of the existing system comes a new spark of hope. Every new experiment in community self-reliance, cooperative organization, worker self-management or localized trade by barter contains the seed of a new social order, and is an important step on the journey toward a truly ecological culture.

6.
Toward a World of Peace and Nonviolence

Probably the most important factor in the rise of the European Green movements in the early 1980's was their uncompromising stand against the stationing of new American nuclear missiles on European soil. These missiles, the Cruise and the Pershing II, were widely perceived by people in West Germany and other countries as a dangerous new destabilizing factor in the international arms race. They were seen as bringing the threat of a nuclear World War III to everybody's doorstep. Since the control of the use of these weapons was to remain squarely in the hands of the United States Air Force, they were believed by many to be a serious threat to European national sovereignty. For many people, this sur-rendering of control seriously compounded the dangers posed by the weapons themselves.

As politicians from across the conventional political spectrum succumbed to political and diplomatic pressures to accept the missiles, the Greens emerged as the voices of a new European peace initiative poised to reject the growing militarization of the world scene by the United States and the Soviet Union. They called for real steps toward disarmament, not just "arms control," and for the creation of an international weapons-free zone that would encompass all of Europe.

Greens also established working ties with the independent, largely underground peace movements in several Eastern bloc countries. They helped organize large demonstrations and campaigns of nonviolent civil disobedience that confronted the institutions supporting the arms race. Greens, both in the European Parliament, and on the streets of Europe's major cities, have spoken against the increasing militarization of the world's cultures and for a new relation of partnership with the people of the Third World.

Most governments in Western Europe have quite willingly accepted their roles in perpetuating the Cold War and supporting the resurgence of militarism, a reality painfully revealed during the Cruise and Pershing debates. However, many Europeans share both a sense of distance from the present centers of power and a fuller understanding of their own war-torn history than is common in the United States today. The European peace movement has evolved a sweeping critique of the historical impulses underlying the global arms race that can help us to better understand our own country's policies and their chilling effects upon the rest of the world.

Greens in Europe have been exploring the logic of "exterminism," the historical unfolding that propels us toward annihilation in the name of security. The British historian and disarmament activist E.P. Thompson pioneered this type of analysis, which seeks to understand just how the development, deployment, and nearly inevitable use of advanced nuclear weaponry has become the driving force of our economy, of technological development and of much of our culture. Never-ending preparations for war, with massive infusions of public funds, sustain the myth of an economy with no limits to growth. Military and corporate planning, centered in the United States, increasingly parallels the highly-militarized economic life of the Soviet Union, with both systems increasingly trapped on a collision course toward mutual extermination. Militarism, in this view, is only in appearance a response to external threats; it more accurately reflects the growing uncertainties of both the American and Soviet economies and political systems.

In today's world, the threat of global war appears increasingly ominous. The United States government is already at war by proxy against the people of Nicaragua and El Salvador. In the Middle East, our government's military posturing, which has little political support outside the United States, is helping to make a highly volatile situation ever-more explosive. The international arms industry, with substantial government backing, is exporting the latest high-powered military technology to all of the world's potential trouble spots. The newest nuclear missiles are increasingly designed to be able to mount a nuclear first strike by virtue of their accuracy and increasingly long range. The initial opposition of European governments to launching a new arms race in outer space has been cynically bought off with offers of new research and development contracts, even though the mobilization of tens of billions of dollars to develop new space-based weapons would represent a qualitatively new step in the militarization of the world economy. Back at home, support for American military aggressiveness appears to be growing, with the help of the news media, the motion picture industry, and all the power and prestige of the White House.

The economy of the United States has grown increasingly dependent on military expenditures to artificially sustain a high level of growth. Many economists describe the federal economic program of the 1980's as "military Keynesianism," a massive increase in deficit spending by the government, diverted entirely to the military sector of the economy. Since the weapons industry is producing no economically-useful goods, free market economists can claim that they are keeping our industrial wheels churning without "distorting" the civilian economy.

This is not a new idea. Several years after the end of World War II, business leaders were concerned that a serious peace initiative by the Soviet Union might destroy the postwar economic boom and increase popular pressure for new public works and welfare programs. A 1949 *Business Week* editorial sheepishly suggested that the coming of peace might force a restructuring of the entire economy as resources spent on the military might become available for other, more popular purposes. The following year saw an unprecedented increase in military spending from $14 billion to $50 billion and a massive increase in American troops both in Europe and Korea.

Today, the Pentagon budget has grown to well over a billion dollars a day. When all of the government expenditures that have military purposes (including payments on old war debts) are added together, they swallow up over two-thirds of all federal tax revenues (excluding Social Security, which is maintained as a separate fund). Weapons manufacturers are guaranteed a fixed profit above their production costs, encouraging the careless waste of resources and skyrocketing costs for ever-more elaborate technologies of destruction. So many regions of the United States are now economically dependent on military industries that even the most liberal members of Congress are reluctant to support meaningful cuts in new weapons programs. Meanwhile, the world becomes an ever-more dangerous place for all of us.

Support for militarism is founded primarily on fear. The Soviet Union plays into Western fears by their continuing Vietnam-style war in Afghanistan. Terrorist groups, especially in the Middle East, increase people's fear by their indiscriminate brutality, religious fanaticism, and merely their differing cultural and social values. In most of today's world, however, the policies of the United States are seen as the major threat to world stability. Our government has blatantly violated international agreements by supporting attacks against Nicaragua and bombing raids on villages in El Salvador. We still provide the major political and economic support for the apartheid regime in South Africa, as well as for police states in Chile, Guatemala, Indonesia, South Korea and other countries. To most outside observers, anti-terrorist posturing by the United States does nothing to relieve the international problems that breed terrorism. Instead, the United States is creating just the kind of militarized political climate, both domestically and internationally, that legitimizes terrorism as a political weapon.

The interests of real global security demand that both the United States and the Soviet Union actively begin to dismantle their bloated military arsenals. Each superpower holds the ability to destroy all life on earth many times over, and the bickering over the details of specific weapons systems that disguises itself as arms control often merely helps both sides modernize their arsenals. The Soviets, much more aware of the strain the arms race places on their domestic economy than most Americans are, have been offering substantial proposals for disarmament through much of the past decade. They suspended nuclear weapons testing for a year and a half, beginning in the summer of 1985, hoping that the United States would respond in kind. Instead, Soviet offers are rejected out of hand, and our arms control agencies continue to be stacked with individuals who have personally opposed most arms control initiatives since World War II.

We in the United States can have little direct impact on Soviet weapons development, except by pressuring our own government to take meaningful steps toward disarmament. These might include a freeze on all new weapons systems, including research and development as well as production, an end to nuclear testing, the disarming and dismantling of nuclear warheads and the elimination of stockpiles of chemical and biological weapons. Arms factories could begin to be retooled and reorganized for making socially-useful goods. How could we better prove to the world that we are serious about peace? American troops and weapons should be withdrawn from Europe, South Korea, the Philippines, Honduras and everywhere else they are poised to intervene in the affairs of other countries.

What Kind of Defense?

A substantial reduction in United States military forces and in the American arms industry would go a long way toward reducing international tensions. However, the political and cultural roots of militarism lie very deep in our history. How are we to assure that a demilitarized America would not be subject to attacks on our own borders? A direct attack is, in fact, unlikely in the present world, as we are separated by vast oceans from every one of the world's other military powers. We are the only country that maintains Rapid Deployment Forces to project our military power overseas with enough force to invade far-away countries without reducing them to useless piles of radioactive rubble. The problem of defending our own borders appears to be a relatively manageable one.*

In Europe, however, the problem of defending one's own borders is much more real. All Europeans lost friends and relatives during World War II and memories of that war's destruction are still quite fresh. This has raised an important debate among European Greens about just what kind of defense is most compatible with Green principles. It is a debate North Americans can learn a great deal from.

One group of Greens, which includes several former NATO

---

*The Coalition for a New Foreign and Military Policy in Washington, D.C. has calculated that only 13% of the 1985 "defense" budget was allocated toward defending the borders of the United States, including those nuclear weapons that are actually poised for retaliation against a possible Soviet strike. If nuclear weapons are omitted, we are left with 3% of the military budget, or a little over $9 billion to defend our own borders.

officers, have developed the concept of a decentralized, non-nuclear border defense. As nuclear weapons are dismantled and foreign troops are withdrawn, each country could line its frontiers with the heavy armor and defensive anti-aircraft weaponry necessary to deter an attack. In different regions of a country, defensive installations would be managed by militias of local volunteers intimately familiar with their region's terrain. All weapons capable of striking other lands and peoples would be dismantled, whether nuclear or conventional. The elimination of large military bases would remove the sorts of targets that invite massive military assaults.

At the same time, we could see a rise in locally-initiated popular diplomacy, helping break the current deadlocks on the road to peace. In the United States, there would be efforts by different regions of our country to engage in pacts of non-aggression and peace with other peoples, including the Soviet Union. One hundred and thirty localities in the United States, and many more in Europe, Japan and the South Pacific, have already voted to declare themselves Nuclear Free Zones. Genuine steps toward economic decentralization would eliminate huge industrial targets, in addition to explicitly military ones. Our "national security" would be based upon the integrity of our own people, rather than the extended web of overseas "security interests" and corporate investments that bind our government to its current interventionist policies.

For many Greens, however, this kind of an approach does not go far enough. They support local disarmament initiatives, but view even a decentralized military defense as a dangerous and shortsighted idea. They feel that it still legitimizes violent solutions to conflicts and does not sufficiently address the cultural roots of militarism. It could create a fortress mentality that sustains rather than dissolves national chauvinism. Even voluntary militias, bringing together and empowering those individuals who have a personal affinity for military weaponry and military ways of being, could actually help preserve our society's most aggressive tendencies.

Greens who profess a personal commitment to nonviolence as a way of life have proposed a different solution. If we are to follow our opposition to the arms race to its conclusion, their argument goes, we should be willing to live our own lives without the false protection afforded by any arsenals of weapons. Violence breeds violence and the ability of any occupying army, or any existing government, to control a population requires that most people passively acquiesce to being ruled. A people united in their desire for freedom and well-prepared in the methods of nonviolent resistance should, in this view, be able to mount a non-military social defense against any potential invader.

Historical precedents for this include the successful resistance of people in the Scandinavian countries to the full consolidation of Nazi control during World War II and the widespread internal sabotage of the Polish economy in the months following the imposition of martial law in 1981. Gandhi and his supporters developed a full-fledged plan to nonviolently prevent a German occupation of India during the Second World War. A high enough level of non-cooperation, civil disobedience and sabotage, it is argued, should be sufficient to render any country ungovernable.* The awareness that a people are prepared to respond in this manner should discourage potential invaders, as the difficulty of mounting a successful occupation would quickly outweigh any possible benefits. Most of the repressive military regimes in the world today could not survive without constant infusions of military supplies and brutal counterinsurgency training, most often supplied by the United States. A Green decentralist outlook, with a firm commitment to personal empowerment through nonviolence, might be the key to breaking the habits of subservience that keep people from fully asserting their freedom.

In the industrialized world today, nonviolent resistance is clearly the most important tool for opposing the arms race and all forms of militarism. Nonviolent action exposes the moral barrenness of the arms race in a society that believes it is committed to peace. Nonviolent activists speak truth to power and help people realize all the ways they quietly support the arms race by simply carrying on their daily lives.

*Many analysts of the military balance in Europe have suggested that the uncertain loyalties of Eastern European troops are a major constraint against the Soviet Union's more aggressive tendencies.

Acts of nonviolent civil disobedience sometimes are chiefly educational in their effect. They can help alert the larger public to a particular issue or injustice by drawing attention to it in a dramatic and creative manner. Sometimes, civil disobedience is useful in helping pressure elected officials to change their votes, a strategy that has met with some success in slowing the escalation of the war in Central America. On other occasions, nonviolent actions can directly impede the functioning of a particular aspect of the arms race, offering a direct example of people power against the military.

In the Western world today, the power of nonviolence is probably best symbolized by the Women's Peace Camp at Greenham Common in Great Britain. Since 1981, women have been camping along the perimeter of the Greenham Air Force Base, the first place where American Cruise missiles were deployed in Britain. The women wander freely around the base, leaving signs of their ability to breach closely guarded areas; every attempt to permanently remove them has failed. On several occasions, tens of thousands of women from all over Europe have participated in demonstrations at Greenham. It is an example of direct nonviolent resistance to the arms race that has inspired similar efforts around the world. One woman, Juliet Nelson, described her arrival at Greenham in the spring of 1983:

> ". . . it was all centered around those missile silos. I think they're a focal point of all the negative things that are going on in the world—paranoia, greed, misuse of power, violence, a lack of imagination for alternatives. In my mind I saw them as revolting man-made boils on the earth's surface, full of evil. I wanted to let out all the feelings I have about the threat of nuclear war—the fear and the dread. And I wanted to concentrate on the future, to feel optimistic and get strength and hope that we can stop it."
> —from *Greenham Women Everywhere*

The Greenham Common Peace Camp has also significantly slowed the deployment of the Cruise missiles, which have been seen as a major destabilizing factor in the arms race. Cruise missiles are not meant to be fired directly from the base; rather they are sent roaming the countryside in "secret" truck convoys. This makes every stretch of roadway within two hundred miles a potential launch site, and thus a potential target for incoming missiles. The presence of the Peace Camp at Greenham has effectively impaired the testing of this mobile launch system. Every time trucks carrying cruise missiles leave the base, a signal is broadcast to peace groups and women's groups all around the world. Local people, fearful for the future of their homes, have set up checkpoints and supported blockades along the convoys' routes. Windshields have been painted to prevent the convoys from proceeding. Many cruise missile tests have been delayed or cancelled.

> On their scale, using their rules, I am powerless. If I try to use their means they will always manage to harm me more than I can dent them. I wouldn't be taking action like entering the high security area of the base and dancing on the half-built cruise missile silos as we did on New Year's Day if I didn't feel frightened and angry about what they are doing. But my anger and fear have to be channelled into creative opposition.
> —Rebecca Johnson, in *Greenham Women Everywhere*

Similar long-term encampments at American military bases have been organized throughout Western Europe. Closer to home, tactics of nonviolent intervention (though involving smaller numbers of people) are being used by the ecological action group Greenpeace to obstruct test detonations of nuclear warheads in the deserts of Nevada and in the South Pacific. In Nevada, they have been joined by peace activists from across the Western United States. Many weapons manufacturers and military bases have been targets of local protests and campaigns aimed at converting them to peaceful uses. Groups of Christian peace activists have been entering large nuclear weapons facilities and hammering on nuclear warhead components in a dramatic enactment of the biblical mandate to beat swords into plowshares. Peace groups committed to direct nonviolent action of all kinds have arisen throughout the Western world as the threat of nuclear war looms closer.

Participating in nonviolent action can help develop the sense of personal empowerment that is needed to create a different kind of world. Whether actions are aimed against the injustices of the present system, or toward creating new ways to defend one's own bioregional community, the practice of nonviolence can overturn the isolation and despair that keeps people feeling powerless. When actions are carried out using a decentralized, affinity group-centered approach, they allow people to rehearse the ways of being an ecological future promises.

Toward a World Without War

War has always been an ecological problem. The earliest warrior societies were also the first that sought to debase and possess nature. The destructive underlying character of much of our technology betrays its military origins, from bronze plated armor to space weapons and new innovations in computer technology.** Recently, it has become apparent that even a "limited" nuclear war—if there could ever be such a thing—might stir up enough dust to create a Nuclear Winter, a protracted artificial winter that would be devastating to much of life on earth.

Radioactive wastelands have already been created by the testing of nuclear weapons in the South Pacific and in remote stretches of Siberian tundra. Modern chemical and biological weapons threaten to render large areas of the earth unable to sustain life. Even so-called conventional weapons, in their newest versions, are so accurate and so overwhelming in their destructive power that even the dismantling of every nuclear warhead would no longer be enough to make the world safe.

The creation and continual replenishment of large arsenals of weapons also has major ecological consequences. Vast wilderness areas, largely in the Third World and in the frozen Arctic, are being mined to sustain the vast stockpiles of "strategic minerals" needed to sustain the American and Soviet military machines. The manufacture of advanced military hardware requires far larger supplies of minerals such as chromium, platinum and molybdenum than are needed even for sophisticated civilian technologies. Many of these minerals come from parts of the world, such as southern Africa,

---

**The Congressional Office of Technology Assessment reported that, in 1983, 87% of the applied computer science research in the United States was paid for by the Defense Department.

where large-scale mining has always been an instrument of colonial power. The bloated resource needs of the international weapons industry have thus created another obstacle to the self-determination of Third World peoples.

Industrial societies, whether supported by a mercantile, capitalist or state socialist economy, have always depended upon the exploitation of faraway lands and people for their survival. Originally, goods were taken from the New World of the Americas to fuel the industrial machinery of England, France and Spain. Today, the international industrial economy thrives on the cheap labor and the unrestrained mining of resources in what we call the Third World. The "free trade" policies that our country's representatives advocate in international forums barely disguise the intent of U.S.-based business interests to control the economics and politics of Third World countries without any constraints.

In the past few decades, people in the Third World have become increasingly vocal about their dissatisfaction with this state of affairs. At the same time, the superpowers have become ever-more rigid in their unwillingness to let Third World peoples evolve an independent course. This conflict, with roots dating back to the very origins of European colonialism, has escalated to an increasingly dangerous pitch in Central America, the Middle East, Southern Africa—every one of our world's most volatile areas. The increasing brutality of the two superpowers in supressing moves toward Third World independence underlies the catastrophic events that make World War III feel ever closer.

Greens in Europe and the United States have generally expressed support for indigenous movements for national independence in the Third World. Many such movements have expressed strong desires for political non-alignment in their earlier stages, but have found themselves increasingly entrapped in the machinations of Cold War politics. In Vietnam, an independence movement that once proclaimed roots in the American political tradition became the target of a United States-directed war of counterinsurgency. The architects of that war explicitly hoped to draw out the most authoritarian and pro-Soviet elements in the Vietnamese opposition. Analyses of the Pentagon Papers, the United States military's own accounts of the Vietnam War, have revealed an effort to prove at all costs the self-fulfilling claim that Third World revolutions could not maintain an independent course and improve the lives of a country's people. In Nicaragua in the eighties, the United States has pursued a similar policy, much to the detriment of the original diversity and pluralism of the Nicaraguan revolution. Still, peace activists and Greens from all over the world have been working to aid Nicaragua's efforts to create a more just society, sending material aid, joining work brigades and sometimes organizing to support specific efforts at social and ecological reconstruction. Whether or not one agrees with the political thrust of any particular revolutionary movement, a Green outlook strongly affirms the right of all people to determine their own destiny. Greens in the United States have begun to search for ways to address Third World issues in a uniquely Green manner.

The role of guerrilla armies in most Third World movements raises a particular dilemma for Greens personally committed to nonviolence. Guerrilla struggles have emerged in response to long histories of brutality imposed upon colonized peoples. Often they are the only means of expression remaining for people whose will to self-determination has been ruthlessly suppressed for generations. When guerrilla-backed revolutions succeed, however, a whole new set of problems can rise to the surface. People generally need to stay armed to protect the gains they have won, and so militarism can remain a dominant cultural and political force even after the necessity has passed. The challenge remains for activists to find meaningful ways to support the gains of Third World people without sacrificing their willingness to criticize the ideological and political excesses of Third World governments.

The world can only move toward peace if we begin to completely redefine the relationship between the northern industrial powers and our "less developed" neighbors to the south. The old colonial relationships need to be replaced by relations of partnership among peoples. Many of the present ruling elites within the Third World are themselves obstacles to such a transformation. Educated in the United States, the Soviet Union or in Europe, they often embrace the idea of rapid development, with all its attendant social and ecological disruptions. This small educated minority is often able to enrich itself while most people's living conditions and basic means of survival continue to deteriorate.

Many of the most striking violations of the land rights of native peoples are in fact occurring within the borders of Third World countries. Governments of both the Right and Left are eager to tap undeveloped lands for their mineral and forest resources and are increasingly doing so at the expense of indigenous peoples. The people of Western New Guinea (colonized by Indonesia) and the Amazon jungles of Brazil, the Chittagong highlands people of Bangladesh, the Mayans of Guatemala, the Miskitos of Nicaragua and many others have been victims of forced relocations, armed attacks and various assaults on their cultural integrity. Representatives of indigenous people on all six continents have called for a new Fourth World consciousness in opposition to the pro-development policies of First, Second and Third World nations.

Many Green ideas about community self-reliance, appropriate technology and indigenous agriculture come from observing traditional Third and Fourth World patterns. E.F. Schumacher developed his idea of a Buddhist economics while participating in rural development work in India. Rather than uprooting traditional patterns, we should look to those peoples who have not yet been thoroughly colonized for clues as to how we can live better on the earth. If technology has a constructive role to play in Third World countries, it is in undoing the ecological damage that our system has already created.

Local self-help projects, like those initiated by the British-based relief organization, Oxfam and other such groups, are one step. Greens can encourage direct ties between cooperative movements in our own country and emerging producer co-ops in the Third and Fourth World as a way of combatting the injustices of the current world market and supporting more democratic alternatives.

One thing we North Americans cannot do is expect other people to retain their traditional ways while we live off the exploitation of their resources. Patterns of domination will only be broken as we begin to change our own way of life. The diversion of an ever-larger share of resources to military use only helps reveal what many European Greens have understood quite clearly—that new military technologies are merely the most overt manifestation of a fundamentally life-denying industrial technology.

In the long-run, creating a peaceful world will require a thoroughgoing transformation of both of the world's industrial systems. We will very likely need to see a contraction of the industrial economies before they can evolve decentralized structures for genuinely satisfying human needs. The Green movement, with its origins in both the ecology and peace movements, has expressed the urgency of exploring the broader systemic changes that can underly a transition to a world at peace.

In a 1981 essay addressed to the European peace movement as a whole, Rudolph Bahro described the challenge ahead. "The whole social organism is riddled by the disease of militarism," he wrote, "and just as it seems that cancer can only be cured at the level of the organism as a whole, so we cannot hope to root out militarism . . . without a similar holistic therapy." His therapy would involve the peaceful dismantling of much of our industrial megamachine, in the hope of preventing a much more catastrophic and disastrous collapse. As the debate over the deployment of Pershing missiles in Germany came to a head, Bahro implored, "we must begin to dismantle the Tower of Babel before it collapses on top of us." We would do well to heed his words.

III   PROSPECTS

A Parable

For thousands of years, the Hopi Indians and their immediate ancestors have lived in peace among the hills and mesas of the desert Southwest, in the land we know as northeastern Arizona. They live a highly spiritual life, with rituals and ceremonies evolved to help them nurture the delicate natural balances that make human life possible in this arid land. Their villages, built into the sides of imposing cliffs, are each independent entities, thriving under the guidance of their traditional spiritual leaders. The very idea of central government is, to the traditional Hopi, a violation of the patterns of nature.

Beginning in the late nineteenth century, children of Hopi families were often kidnapped by missionaries of the Mormon church, who taught that the red people were an impure race and that the material resources of Mother Earth were there to be taken for the benefit of human progress. A sect of "progressive" Hopi, who had internalized their Mormon training, emerged and were established in the recent years as officials of a "tribal" government with the support of the United States. They signed leases with Utah-based mining companies, who sought access to the huge deposits of coal and other minerals that lay beneath Hopi lands. They conspired to legislate the removal of thousands of Navajo people who had come to settle around the perimeter of Hopi Land over several hundred years, people who had evolved their own understanding of the sacredness and inviolability of the desert mesas and had served for generations as a buffer between the traditional Hopi villages and the earth-denying, power-hungry civilization surrounding both nations. The effort to prevent this forced removal has recently brought together people from across the continent who share a reverence for traditional native ways.

In our time, the elders of the traditional Hopi have chosen to make known a prophesy that was carved in stone by their ancestors outside the ancient village of Oraibi. It depicts a Time of Purification, when red and white brothers would reunite, when the corrupted people, who think only with their minds, would rediscover the natural way, the Way of Peace. This would follow a time of great restlessness and dislocation, of the disruption of the seasons, of the chemical manipulation of animals and people, of communication through cobwebs (telephone lines), of the symbols of the cross, the swastika and the sun (symbol of the imperial Japanese), and of the fiery gourd of ashes (the nuclear bomb) that would spread incurable sickness among humankind.

The traditional Hopi bring a message of hope and renewal in these disturbing times. Whether we accept their image of a personified Creator delivering wisdom to the people, or believe spiritual forces to be emergent from nature itself, we can all gain inspiration from the power of their message and from their ability to sustain their traditional ways amidst unfathomable pressures to conform to the acquisitive, materialistic society that now surrounds them.

Grandfather David Monongye is the spiritual leader of the independent Hopi village of Hotevilla. His words call out to all of us:

> Perhaps there is still time left to reawaken the misguided and prevent disaster. We learned from our ancestors that man's [sic] actions through prayer are so powerful that they decide the future of life on earth. We can choose whether the great cycles of nature will bring forth prosperity or disaster. This power was practiced long ago, when our spiritual thoughts were one. Will this concept still work in the jet age?
>
> Let us not be discouraged. Let us cleanse our minds of delusion. Let us rid ourselves willingly of hate, and put love within ourselves, and join together with renewed faith in our Creator, so that we may be spared the destruction that results from

trusting in weapons and other devices of our own minds, and not forget the future of our children and those to come.*

The choice between purification and destruction is truly ours. This time, we can be certain that the whole world is watching.

*From "A Message to the Dalai Lama of Tibet," October 1982

7.
How Can We Create a Green Future?

The Green vision of a peaceful world moving beyond industrialism is changing the terms of political debate in much of Europe today. As governments throughout the West succumb to Cold War pressures and endorse policies that make the world an ever-more fearful place, Greens are offering a vision of a different kind of society. Green movements in many countries have tapped a deep-seated discontent with the present state of affairs that reaches across the conventional political spectrum. They are striking a vital chord of hope in a time of increasing despair.

The German people's traditional love for their land, once ruthlessly manipulated by Nazi leaders, has provided fertile ground for this new vision of freedom in a world without weapons. Today, people from Scandinavia to the shores of the Mediterranean, from the British Isles to the Danube and eastward are creating the foundations for a Green Europe, free of the militarism and the economic pressures imposed by the two superpowers.

From its European origins, Green politics is spreading around the world. Australia and New Zealand, each with their own growing ecological awareness and a heritage of political independence, have become centers of resistance to nuclear technology in all its forms. In Japan, today's leading symbol of technological achievement, a Green movement has developed, with a strong awareness of the links between political and personal changes. Greens are appearing in the Third World, too—in Brazil, several long-time activists, including some prominent former guerrillas, have formed a Green party dedicated to nonviolence, racial and sexual equality and political independence.

134

Can it happen here? Can a new ecological politics take hold in the United States, where the frontier myth of prosperity at the expense of nature reigns supreme? The opinion-shapers in the mass media tell us that the age of discontent that began in the 1960's is over, that we have entered a new era of complacency, materialism and an acceptance of selfish, competitive values. People are no longer interested in changing the world, we are told, just in getting ahead and making a comfortable life for themselves, whatever the consequences.

But for many people, a high quality of life demands acknowledging one's responsibility to the earth and to other peoples. The spread of nuclear weapons, the African famine and the nuclear meltdown in the Soviet Union have all brought this awareness to the surface. Although most people seem to quietly accept our government's bullying stance toward the rest of the world, there is also a healthy mistrust of powerful institutions that was not apparent a generation ago.

Our society is deeply polarized. Behind a veil of prosperity and economic recovery, new economic pressures are afflicting both the urban and rural poor. The affluent rush to gentrify inner city neighborhoods and buy new vacation homes in the countryside, while increasing numbers of long-time farm families and middle class homeowners are facing foreclosure. Consumerism appears to be on the rise, while more Americans than ever before are going hungry and the ranks of the homeless are swelling at an astonishing pace. While some people are seeking a gentler way of life and a healed relationship with the earth, many of their peers are finding newly-fashionable rationalizations for joining the culture of exploitation and greed. People confronted with their own isolation in an increasingly insecure and atomized society are abandoning earlier dreams of community to pursue their own personal fortunes. This new cult of selfishness encourages support for overt militarism and nationalistic jingoism, which often exist right alongside the skepticism toward institutions that came to the fore during the sixties.

It is a time of searching and a time of vision. A new spiritual yearning has reached segments of our society that seemed unaffected by the changes of the sixties and seventies. Many people are seeking a new spiritual underpinning for their lives. Some seek the comforts of traditional religion, and our country's churches and synagogues are increasingly reflective of the moral tensions of our time. From the rise of liberation theology and the nationwide sanctuary movement to protect Central American war refugees, to the new politics of fear being spread through the fundamentalist churches, religion and politics are no longer as easily separable as we used to think.

In this highly-charged climate, conventional politics are increasingly suspect. Traditional liberalism is largely discredited, as the moral fervor it appeared to command in the early 1960's has given way to reveal a managerial politics that seeks shallow, bureaucratic solutions to deep-seated social problems. Conservatives claim a moral upper hand today, but a look below the surface of this modern conservatism reveals its subservience to corporate power and its profound disrespect for individual rights. A moral vacuum has developed, one which the ideologists and television preachers of the far right have rushed to fill with their own manipulative calls to community and their overflowing confidence in the repressive values they call "American." A new right-wing establishment consolidates its power with populist appeals to defy "the establishment."*

The ecological crisis and the threat of nuclear war add a heightened sense of urgency to the situation. In the face of unprecedented threats to our species' survival, it is too late to expect a solution to emerge from anywhere on the conventional political spectrum. It has become necessary for peace-loving and ecologically-minded people to articulate not just a new politics, but a new ethics and a new earth-centered moral sensibility that can reawaken the life-affirming impulses our society seeks to submerge.

This new ecological sensibility is emerging in a variety of ways, from many sectors of our society. The preceding pages have

---

*For all their talk about community, freedom and "traditional" values, the Right's political agenda seeks to limit individual rights and police people's private lives, while political constraints on corporations are increasingly relaxed. *Economic* activity, for them, is to be free to expand without restriction, but the rights of people and communities to protect themselves are increasingly restricted from above.

touched on just a few of these. Bioregionalists, eco-feminists, community activists, spiritual healers, earth-guided poets and curious people of all walks of life are raising hope for the evolution of new attitudes and new ways of relating to the earth and to our fellow beings.

The Green movement represents an active expression of these values. We have seen how Greens are beginning to fit the pieces together, discovering how ecological principles can guide us on a path toward the necessary social transformation. But many questions remain. How can the various ecological currents evolve together to create a dynamic whole that is greater than the sum of its individual parts? How does a Green outlook change the ways people work for social change? How can local autonomy be preserved and strengthened as Green values are articulated and expressed on a regional and national scale? When is it appropriate for Greens to participate in conventional politics? What is the relationship among political, social and personal changes?

Greens in the United States are working today in many different spheres and using many different approaches. Some people are focusing their attention on specific issues of local and regional concern. They are working to curb the excesses of industrialism and to head off the social and ecological disruptions rooted in our present way of life. This current is basically oppositional in nature, embracing political methods such as community organizing, lobbying for legislative reforms and a variety of direct nonviolent efforts to protest or obstruct especially threatening policies and projects.

The second major current is reconstructive in its approach. It includes a wide variety of efforts to create living alternatives to our present ways—a wealth of experiments in cooperation and local democracy—both in the community and the workplace. It includes the development of alternative technologies and the raising of bioregional awareness. For a Green movement to develop and grow in North America, there will need to be a merging of oppositional and reconstructive strategies that allows these two currents to support and strengthen each other.

Issue-oriented politics without an alternative vision can be politically limiting and personally frustrating. Many people are uncomfortable with the way things are, but are not motivated to act on their beliefs because they see no other way. Others might choose to work on a particular issue of concern, but are easily exhausted as each small victory reveals new complications. One might work for many months to block a particularly devastating project or to achieve a particular reform in the system, only to find that new injustices crept in the back door while your attention was focused on one small piece of the problem. The ecological crisis cannot be simply controlled within the limits of the existing system. In fact, some Greens believe that reformist efforts merely forestall the impending collapse of the industrial economies, a collapse which may need to occur before the real work of reconstruction can begin.

It can be equally limiting to work to create new institutions without actively seeking to understand and oppose the injustices of our present ways. Such efforts can be slowly bought off and accommodated into the service of the present system. One can point to food co-ops that have become more involved in elaborately marketing their goods than in fully challenging the limitations of the existing food supply system. A once-vibrant alternative energy movement in New England has become tied to the ecologically-devastating vacation home industry, as solar builders have drifted toward affluent resort areas in their search for steady employment and the freedom to experiment. Should healthy food and solar-heated homes become the luxury goods of an affluent minority seeking to purchase an ecological "lifestyle"? How can a Green sensibility guide us toward a better way?

The West Germans have borrowed a phrase (originally attributed to the ecologist Rene Dubos) that has become a slogan for the worldwide peace movement: "Think globally, act locally." Local ecological problems, local symbols of the military-industrial complex, and local attempts to create alternatives in housing, food distribution and other basic needs all offer a focus for local activities that carry a global message. By working primarily on the local level, Greens are demonstrating the power of people to really change things and creating the grass roots basis for a real change in consciousness.

In local issue-oriented work, a Green sensibility offers new opportunities to link specific issues together. For example, local weapons industries fuel the arms race, distort employment patterns in a region, and often produce the largest quantities of toxic wastes. Campaigns aimed toward converting such facilities to peaceful uses can raise questions of community and worker control and the nature of economic and technological change. Large development projects often destroy prime farmland, wildlife habitats and people's homes, while centralizing the control of a local economy in fewer hands. A Green anti-development campaign can fully integrate these closely related concerns.

Traditionally, coalitions of different political groups are very narrowly defined. Several organizations come together to deal with a very limited set of common concerns and the coalition's definition is confined to a lowest common denominator of what different groups can agree upon. Green politics suggest a different kind of coalition, in which people are united by a larger vision of an ecologically-transformed society. The Ozark Community Congresses and other bioregional gatherings are examples of this new kind of coalition-building. In San Diego in the 1970's, a coalition of publicly-supported educational and social service organizations created a visionary Community Congress to pool organizational resources and promote the greater empowerment of people in a community setting.

In the environmental movement, a Green outlook offers a new grass-roots focus and more democratic organizational models. As the large national environmental groups become increasingly focused on events and personalities in Washington, D.C., new kinds of groups have emerged to fill in the gaps. The national movement against hazardous waste dumping and toxic chemical pollution is one example of a different kind of approach—toxics activists have primarily devoted their energies to helping people form local groups to confront local problems. National organizations gather and publish information, offer technical help, teach organizing skills and engage in lobbying at the federal level. Their lobbying efforts, however, are backed up by hundreds of large and small citizens' groups organized and ready to take action to protect their communities.

Another new model for environmental politics, one grounded in a more sweeping ecological vision, is offered by the Earth First! organization. The founders of Earth First! (always spelled with a "!") were inspired by author Edward Abbey, whose writings embody the rhythms of the land and people of the Southwestern deserts and prairies. His popular novel, *The Monkey Wrench Gang*, describes the often comic adventures of a band of renegades committed to reversing the commercial exploitation of their beloved desert by all means of disruption and sabotage.

Begun as a loose network of people committed to environmental direct action in the Southwest, Earth First! has grown to include over fifty affiliated groups all across the continent. "Monkey wrench" tactics are still very much a topic for discussion, but the focus of Earth First! has broadened to encompass a wide variety of educational activities, including taking guerrilla theater from city streets to the National Parks. They have developed detailed plans for the reclamation and expansion of wilderness areas threatened by development, carrying their plans to the appropriate officials with the backing of national letter-writing campaigns and, when necessary, through sit-ins, office takeovers and the direct obstruction of logging and road-building operations.

Earth First! groups are not tied to any formal organizational structure, but keep in close contact through the mails and through their newspaper, which has become a major national forum for the renewal of ecological activism and thought. They have also helped publicize action campaigns to save wilderness in other parts of the world—in recent years, people have put their bodies on the line to hold back bulldozers and chain saws in the Australian rain forest, on the last remaining wild stretch of the Danube River in Austria and on Indian lands in British Columbia.

A Green perspective can also change the way people work to create alternatives to our present way of life. A Green vision encourages strong ties between alternative efforts in different spheres, alliances between consumer and producer co-ops, and the adoption of more democratic styles of organization. Rather than simply providing goods and services that may or may not be available through conventional channels, Green efforts at creating alternatives can help people to begin actively withdrawing from the system that oppresses us all. At a time when many former activists are drifting toward more mainstream ways, a Green movement can help put the search for entirely new ways of living and working together, grounded in a traditional closeness to the land, back on the social agenda.

Many of the experiments in community-based economics that we discussed in Chapter 5 are living proof that these changes are occurring. The emergence of both urban and rural Land Trusts is a promising step toward the evolution of closer relationships between people and the land, as well as new economic and social arrangements. Experiments such as those in the Berkshire Mountains of Massachusetts offer new ways for communities to devote local resources to the satisfaction of local needs. Community systems of exchange and barter based upon relationships of trust can begin to free us from some of the limitations and built-in inequities of the dominant economy.

One Green group in the central hills of Vermont has proposed the development of alternative economic experiments in tandem with the creation of a Community Congress of existing community institutions. By merging these two efforts, they hope to encourage a new strength of working relationships among the different sectors of both the alternative and traditional communities. In Boston, some Greens are working with activists in the black community to turn an abandoned state hospital site into a center for education and experiments in urban agriculture and alternative technologies. The center would employ local people, teach new skills necessary for self-reliance and provide healthy food for inner-city residents. At the same time, the project would help curtail land speculation and commercial over-development in the midst of a residential neighborhood by averting plans to turn the site into a new high-technology industrial park.

Greening the Electoral Sphere

As Green ideas become more widely understood in this country, a third sphere of activity becomes increasingly attractive to Green activists, the world of electoral politics. The electoral successes of the European Greens have raised hopes that Greens in this country could directly influence local and national policies by campaigning for and winning elective offices.

In a few locales, Green electoral efforts have already begun to bear fruit. In New Haven, Connecticut, a Green Party was formed in 1985 to run a slate of candidates for the City Council. Greens came in an impressive second to the entrenched Democratic machine in most districts where they fielded candidates, and their mayoral candidate outpolled the Republicans in nearly a third of the city's wards. Issues of housing, commercial development, waste disposal and the tax-exempt status of Yale University gained new importance in the city. One incumbent City Council member joined the Green Party several months after the election, becoming the first elected official in the United States with a Green affiliation.

Greens in Maine, North Carolina, Michigan and elsewhere have become involved in local electoral politics and met with some successes. In California, the Peace and Freedom Party, which was established as an alternative electoral party in the 1960's, has begun to incorporate Green ideas into its statewide campaigns, and several local Green groups have begun to contemplate electoral involvements of their own. Many more such efforts are likely to emerge in the coming years as Greens in other parts of the country choose to bring their ideas to the electoral arena.

Green electoral campaigns can help bring ecological ideas to a wider audience. They can attract substantial media attention and provide an opening for Greens to speak with a greater cross-section of their neighbors. Sometimes these efforts will have a tangible effect on public policies, occasionally as a result of a Green candidate being elected to office, but more often by reshaping public debate around important issues. In a time when the terms of national political debate are so influenced by the extreme right, local Green candidates can help maintain a public focus on issues of environmental protection, democracy and human liberation. When broad popular support for such causes can be demonstrated, it can have a significant impact on the positions taken by major party candidates.

Green electoral politics is not without its serious pitfalls, however. It is one thing to acquire an "audience" for Green ideas; it is quite another to create activities that really change political institutions. Electoral involvements can complement more grass-roots efforts, but they can also undermine them by helping sustain the illusion that we can change the way things are by simply choosing the right leaders to represent us.

The rural state of Vermont has been a center of alternative electoral politics since the early 1970's. Its small towns, face-to-face institutions and its part-time citizen legislature have encouraged people outside the mainstream to become involved in local politics. Vermont's largest city is Burlington, a booming commercial center with a pace of political activity uncharacteristic of a community of only 38,000 people. The electoral successes of Burlington's progressive-left community offer many lessons for Greens and others considering an electoral approach to social change.

When Bernard Sanders, an independent socialist, was elected mayor of Burlington in 1981, many people active in local efforts for peace and social justice chose to put other priorities aside for a time to support his administration. Sanders promised a voice in government for those who had been disenfranchised by the manipulations of corporate power and pledged to stand up to the powerful financial interests that had controlled the city for decades. But from the beginning, much of the focus was upon Sanders' personality, his rebel image and his charismatic oratory. Supporting that image and defending the mayor in his various personal battles with the entrenched city bureaucracy became an increasing preoccupation of local activists.

In the neighborhoods of Burlington, people got the message that they finally had a mayor who would fight their battles for them. Community activists newly appointed to administrative posts in City Hall were far more willing to listen to people's grievances than their predecessors. The various neighborhood-based organizations that might have become the foundation of political activity in Burlington largely fell by the wayside as both the activists and their constituents focused their attention on the mayor and his priorities.

The trading of election news and gossip became a growing preoccupation, functioning almost as a substitute for grass roots activity. Meanwhile, the local bankers and real estate developers continued to build up and gentrify the city at a phenomenal pace. Sanders struck compromises with them to attract jobs and to soften their electoral opposition. He attacked peace activists for challenging a local weapons plant and offered mere lip service to environmental and feminist concerns. Still, several years were to pass before the glamour wore off and community activism in Burlington began to reassert itself, with a renewed focus on housing issues and the problems of over-development. Greens played a leading role in this re-emergence through their opposition to a multi-million dollar waterfront development that was actively supported by the Sanders administration.

A similar predicament could easily arise in a community that succeeded in electing Greens to public office. Our political system offers little breathing room for the kind of active relationship between elected officials and grass roots efforts that Greens in Europe strive for. The dominance in American political life of the mass media makes the situation even more difficult, especially in urban areas. Once people experience the rush of media attention that comes with electoral campaigns, it seems difficult for them to return to slower-paced community work.

As Green ideas spread, politicians are likely to come forward who will claim to speak for ecology and for a politics of decentralism and home rule. Such individuals will probably emerge both from within local Green organizations and from conventional political circles. Many will be sincere in their desire to be a voice for this new politics and bring genuinely transformative proposals into the mainstream arena; many will be using Green ideas primarily as a vehicle for their own political ambitions. Before diving in to support such individuals, Greens should ask whether such an effort will really enhance the cause of popular empowerment. One possible litmus test might be the potential candidate's willingness to subordinate their personal ambitions to the needs of local Green alliances that are fully democratic and open to wide participation. It is important to maintain the distinction between individual political campaigns and efforts to create a political expression for a larger ecological and social movement.

The uncertainties of Green electoral politics in the United States are compounded by the important differences between the American electoral system and those of most of Western Europe. In Chapter 2, we saw how the European systems of proportional representation allow electoral minorities to gain a parliamentary voice. Parties representing a fairly wide spectrum of ideas can participate in elections and expect to voice their own views, however controversial, in their countries' legislative bodies.

In the United States, as in most English-speaking countries, the winner-take-all approach to elections has effectively prevented a full spectrum of political parties from emerging. There is no basis for a debate about the nature of political compromise, since only candidates who purportedly express a "majority" view have a real chance of getting elected. Political principles are subordinated to displays of personality as campaigns are increasingly fought over images manufactured by and for television. In highly-charged times such as ours, real differences in values can sometimes rise to the surface during election season. However, our electoral system compels politicians of both major parties to behave very much alike once they are in office. Most people feel they have no recourse other than to wait until the next election and vote for somebody else.

Green politics raises new hopes for finding a way to engage the system from within and at the same time create alternatives to it. Can Greens attempt this without becoming like other politicians? First, it will be necessary to dissolve the traditional division between leaders and followers, which is so heavily reinforced by the mass media. Campaigns will need to help raise people's confidence in their own power to change things. Greens will need to discover ways of speaking to mainstream Americans without compromising basic ecological principles. Candidates and their supporters will need to serve as organizers and catalysts for grass roots activity rather than as just another crew of packaged personalities and salespeople.

In Maine, Greens first came together in 1984 as a statewide political party. They hoped to sustain the electoral machinery of a nearly-successful referendum campaign to close the state's only nuclear power plant. But several months of organizational difficulties convinced most activists that a more decentralized approach was needed. Now, several county-based Green groups across Maine are carrying out a wide variety of ecological campaigns, educational projects and local electoral efforts.

The New Haven Greens also formed around an electoral campaign, but soon came to see their electoral work as just one part of a larger consciousness-raising effort. They felt their neighbors wanted to see them first challenge the city's political machine, but a broad range of neighborhood concerns have since come to the fore. The openness of their campaign organization—strategies were decided weekly at open neighborhood meetings—and their use of guerrilla theater and other forms of public protest helped set the stage for their ongoing work. A Green outlook, when carried to its full implications, can help transform politics from the distasteful spectacle it has become, to a vehicle for asserting popular control and popular empowerment.

In the short-term, Green electoral campaigns could help mobilize an ecological constituency that politicians would then have to pay attention to. They could help push the system to reform itself, as established institutions seek to accommodate popular movements and mainstream politicians try to catch up with their constituents' views. But until we begin to see real changes in our political institutions, all types of ecologists are easily lumped together as just another "special interest" whose demands simply get in the way of corporate and other elite efforts to manage our society.

In the middle 1970's, a group of influential business leaders from the United States, Western Europe and Japan formed a panel called the Trilateral Commission to begin formulating economic and political strategies for the decades to come. They commissioned a study of the domestic climate in each of the three "trilateral" regions, a study which appeared in 1975 under the title, *The Crisis of Democracy*. In the United States, they concluded that a "democratic distemper" had arisen during the 1960's, that too many new sectors of the population had come to expect a voice in public decision-making. This "excess of democracy" had, for the Trilateralists, led to a problematic rise in government activity and a decline in its authority. It was deemed essential to return to the climate of the early fifties, when there was little challenge to the dictation of economic and foreign policy by powerful business interests. A revival of charismatic leadership and institutional power would be necessary, they predicted, to return the American people to a sufficiently passive state.

The 1980's saw the fulfillment of their predictions to a degree few would have expected. Even though most Trilateralists considered themselves to be of a more liberal persuasion than the Reagan Republicans, there would be little disagreement between the two groups about the threat a politically-active citizenry poses to the dominance of elite institutions over political life. People's desire to control their own lives creates a real dilemma for those who have long been able to manage society to serve their own economic and political ends. The radical decentralist outlook of many Greens escalates this dilemma one step further: if federations of self-reliant communities begin to take meaningful steps toward withdrawing from the system, what measures will be taken to prevent the "distemper" from spreading even further?

A major hope for Greens everywhere lies in the development of new community-based institutions and experiments in local democracy. Such efforts could begin to create a genuine counter-power to the influence of established institutions. If a few communities in a few regions can begin mapping out a more independent course for themselves, they can help others discover how to break the web of dependencies that keep people believing in the present system. As the system increasingly fails to satisfy many people's most basic needs, the search for alternatives can evolve to an entirely new level.

In the near-term, the Green movement can become an important vehicle for renewing public activity at the local level. It can articulate a sweeping critique of our present way of life and catalyze efforts to develop bioregional consciousness and an ethic of local self-reliance. Green groups in neighboring communities can federate together to create changes in economic and social policies on a wider scale. We can begin at the grass-roots level to transform our relationships with peoples in other parts of the world. A new politics of neighborhood and community, grounded in an ecological awareness, can offer an essential challenge to the forces of destruction at work all around us, a challenge that can spread by example to all the far corners of our land.

Beyond Politics

The success of Green initiatives, both in community organizing and in the electoral sphere, raises new hope that an ecological transformation of society is indeed possible. A Green outlook, combining social and ecological responsibility with a commitment to democracy, nonviolence and a thorough decentralization of political and economic power, provides a framework for addressing immediate concerns and for taking steps toward a different kind of future. But just as a society's relationship to nature is shaped by all of that society's social, economic and political institutions, the structure of these institutions is shaped by the nature of people's relationships to them and to each other.

In Chapter 1, we considered some of the steps people are taking on an individual level to heal their own personal ties to the natural world. It was suggested that personal changes are incomplete without broader changes in society. Now, the discussion comes full circle, as we begin to see how changes in the political structure of society are also insufficient by themselves. A Green perspective addresses the need for changes on many different levels, acknowledging the complex relationship between the political, social and cultural spheres. As Green activists work to change political structures and institutions, it is equally important that the foundations of an earth-centered culture, a personal ethic of cooperation and new ways of expressing our intimacy with the natural world are brought into being.

An ecological outlook proclaiming a renewed harmony between humanity and nature can also be an inspiration for a new harmony among the political, social and personal spheres within human communities. As Green politics seeks the enhancement of human diversity and human freedom as essential steps toward healing our relationship to nature, it can point toward new ways for the often-conflicting spheres of human activity to better complement and nurture each other.*

In a world where "politics," for most people, means the proclamations of politicians and the petty squabbles among government officials, Greens promise to give the word a new meaning. Green politics embodies a new understanding of the public sphere as a forum for enhancing the living interrelationships among people and communities, not just a different way of administering the ins titutions of the state.

A Green view of society calls for the full expression of people's individuality within a community setting. Society, for many people,

---

*The Green commitment to freedom and social justice in human communities distinguishes it from ideologies that invoke "natural law" as a limitation on human possibilities and as a constraint on reason and creative thought. The Nazis, for example, evoked a caricatured reverence for nature that was merely a rationalization for their murderous racial theories.

has become a vehicle limiting their individual expression, a realm of conformity and enforced social roles. Many people are still fleeing the provincialism of traditional communities, even while others are discovering anew the deeply-felt need for community that urban society leaves largely unsatisfied. An ecological model of unity-in-diversity and a renewed ethic of cooperation among people can help transcend this dilemma. A social realm informed by an ecological wisdom, striving for self-reliance at the community level and freed of the acquisitive pressures of contemporary mass culture can become a true vehicle for personal liberation.

A Green perspective also calls for a deepened personal relationship with the natural world. A reawakened awareness of the wholeness of nature can help enhance a person's own sense of self and enrich one's deepest personal relationships. The concept of individuality, in a life-denying culture, often implies a hostile, competitive relationship toward one's fellow beings. From an ecological viewpoint, every aspect of one's personal life can come to reflect a sense of celebration and connectedness with the natural world, and with the rest of humanity as well.

The Green movement itself offers some important tools for reshaping the ties between the political, social and personal spheres. A consensual approach to working together in groups validates everyone's unique contribution to the creation of a different world. It brings the hope of building trusting, cooperative relationships among people right into the structures created for working together. The goal of consensus also carries a commitment to seeking cooperative, nonviolent methods for resolving and mediating disputes. Similarly, a confederal approach to social change, with broader changes in society emerging outward from the local level, affirms the importance of individual community efforts. Every activity that plants a seed of transformation in one area helps contradict the dominant culture of passivity and subservience, opening up new channels for cooperation among people and among communities.

Finally, Green political groups can help nurture the personal development of their members, spreading an awareness of the personal impacts of political events and the political consequences of one's personal choices. Green political events can embody a genuine spirit of celebration, enfolding people's deepest artistic and spiritual expressions into the work of changing society. An ecological movement thus fully values people's contributions in all fields of endeavor and all forms of expression.

All across the land, the basis for a new earth-centered culture is being cast. Artists and poets are turning back to the earth as a source of inspiration. Naturalists and philosophers are bringing a new sense of "deep ecology" into their work. People are creating new kinds of personal and community rituals to express their bond to the earth, viewed once again as the nurturing mother of all life. Even in the natural sciences, the ancient view of the earth as a living being is inspiring new understandings of the regulation of the atmosphere, the origins of life and the role of cooperation and of consciousness in biological evolution.

However, we are still living in troubled times. The political Right succeeded to a truly shocking degree in using their command over national policy in the 1980's to cultivate disturbing changes in people's attitudes and aspirations. They have helped raise to the surface some of the greediest and most chauvinistic elements in the American national psyche. In this time of change, many people are withdrawing like squirrels in autumn to protect whatever resources they can claim for themselves.

In the emerging Green movement, we can see the beginning of a different way. We find people openly embracing a need for community and an ethic of cooperation. We meet a reactivated citizenry standing against war and injustice, and rebuilding a peaceful society from the ground up. We discover social change work that is integrated into the daily life of communities all across our continent, spreading its message by example. We feel the birth pangs of a new culture of empowerment and a new politics of celebration.

NOTES ON SOURCES

I have attempted to compile a listing of the most comprehensive and accessible sources on most of the topics presented in *The Green Alternative*. The following notes also list the sources of all quotes, statistics, and material given special emphasis in these pages. I have also cited, as sources of further background material, several books and authors that have been especially important in my own personal development, as well as the addresses of organizations that are involved in ongoing political work around issues and concerns that are central to the Green vision.

Introduction

A highly readable account of the West German Greens, though from a distinctly North American perspective, is *Green Politics*, by Charlene Spretnak and Fritjof Capra (Revised edition, Bear and Co., Santa Fe, 1986). Many Green groups in the United States point to this book as a source of their initial founding inspiration. Articles by Carl Boggs and John Ely in *Radical America*, Vol. 17, No. 1, January, 1983 (38 Union Square, Somerville, Mass. 02143) offer a different analysis of the Greens' origins. There are three firsthand accounts by German Greens available in English: *From Red to Green*, a series of interviews with Rudolph Bahro in which he describes his own fascinating political history (Verso/New Left Books, London, 1984); a more recent collection of Bahro's essays, *Building the Green Movement* (New Society Publishers, Philadelphia, 1986); and a translation of Petra Kelly's *Fighting for Hope* (South End Press, Boston, 1984). From a British perspective, there is *Seeing Green* by Jonathon Porritt (Basil Blackwell, New York, 1985).

The outlook of natural ecology is explained in great detail in the

classic textbook by Eugene P. Odum, *The Fundamentals of Ecology.*
Chapter 7 contains a provocative discussion of competition and
cooperation in ecosystems. The vision of social liberation based on
ecological principles that I espouse in these pages is greatly
influenced by the social-ecological writings of Murray Bookchin,
especially *The Ecology of Freedom* (Cheshire Books, Palo Alto, Calif.,
1982).

Renewing the relationship between humanity and the rest of
nature is the theme of an anthology, *Deep Ecology*, edited by Michael
Tobias (Avant Books, San Diego, 1984) and a collection of essays by
Bill Devall and George Sessions, also called *Deep Ecology*
(Peregrine/Smith, Layton, Utah, 1984). Deep ecology is a phrase
coined by the Norwegian philosopher Arne Naess (represented in
the Tobias anthology), whose writings have inspired a heightened
interest in environmental ethics and ecophilosophy.

1. We Are All Part of Nature

My descriptions of the lives and world view of tribal peoples are
based on those in *Stone Age Economics,* by Marshall Sahlins (Aldine-
Atherton, New York, 1972), *Society Against the State* by Pierre
Clastres (Urizen Books, N.Y., 1977), *The Underside of History,* by Elise
Boulding (Westview Press, Boulder, 1976) *The Phenomenon of Life,*
by Hans Jonas (University of Chicago Press, 1982) and *Mutual Aid,*
by Peter Kropotkin (Porter Sargent, Boston).

The Mohawk Nation's *Basic Call to Consciousness* (1978), published by Akwesasne Notes (Mohawk Nation, via Rooseveltown,
New York 13683) is the source of most of the Native American
quotes. The quote from Dhyani Ywahoo is from an interview in
Woman of Power, (Issue #1, Spring, 1984, available from P.O. Box
827, Cambridge, Mass. 02139). Many more descriptions of Indian
world views, explained in their own words, can be found in *Touch
the Earth,* edited by T.C. McLuhan (Pocket Books, N.Y., 1972). The
question of leisure time in peasant societies is discussed by E.P.
Thompson in "Time, Work-Discipline and Industrial Capitalism" in
the British journal, *Past and Present* (No. 36, December, 1967). The
works of Christopher Hill are also useful for insights on that
period.

Murray Bookchin's *The Ecology of Freedom (op. cit.)* offers a sweeping ecological interpretation of the emergence of hierarchy from within traditional societies. It traces the parallel evolution of social domination and the idea of dominating nature right up to the present day. *The Underside of History,* by Elise Boulding *(op. cit.)* offers additional insights. Some important spiritual roots are explained in Dolores La Chapelle's *Earth Wisdom* (Finn Hill Arts, Silverton, Colorado, 1978).

Lewis Mumford's monumental history of technology and its social roots is in two volumes, *Technics and Human Development* (Harcourt Brace Jovanovich, New York, 1966) and *The Pentagon of Power* (Harcourt Brace Jovanovich, New York, 1970).

On the desecration of Indian lands and cultures, see Frederick Turner's *Beyond Geography* (Rutgers University Press, New Brunswick, N.J., 1983) and *Changes in the Land,* by William Cronon (Hill and Wang, New York, 1983). Turner's book emphasizes the crucial role of Christian ideology in the European conquest of the Americas. A more optimistic view of the role of Christianity in fostering a Green sensibility is offered by Charlene Spretnak in *The Spiritual Dimensions of Green Politics* (Bear and Co., Santa Fe, 1986). An eye-opening account of Christian-inspired utopian movements during the Middle Ages is offered in Chapter 8 of Bookchin's *The Ecology of Freedom (op. cit.)*

For an insightful feminist analysis of the origins of the scientific world view, see Carolyn Merchant, *The Death of Nature: Women, Ecology and the Scientific Revolution* (Harper and Row, San Francisco, 1980). Also see Lynn White, Jr., "The Historical Roots of Our Ecological Crisis" *(Science,* Vol. 155, p. 1203, 1967) and David Kubrin, "Newton's Inside Out!—Magic, Class Struggle and the Rise of Mechanism in the West" (in H. Wolf, ed., *The Analytic Spirit,* Cornell University Press, Ithaca, N.Y. 1981). The Descartes quote is from the *Discourse on Method* (Library of Liberal Arts, 1960). Francis Bacon's fanciful novella, *The New Atlantis,* is available in *Famous Utopias of the Renaissance* (F.R. White, ed., Hendricks House, New York, 1955). Another good introduction to the scientific revolution is H. Butterfield's *The Origins of Modern Science* (Macmillan, New York, 1958). The mystical underside of Renaissance rationalism is revealed in *Giordano Bruno and the Hermetic Tradition,* by Frances Yates (University of Chicago Press, 1964). Susan Griffin's *Woman and Nature* begins with a poetic overview of the misogynist origins of the scientific world picture (Harper Collophon, New York, 1978) and is the source of the quote from Kepler. The Bacon quote is from *The Death of Nature (op. cit.)*. On the peasant roots of the English revolution, see Christopher Hill, *The World Turned Upside Down* (Viking, New York, 1972) and D. Kubrin, (*op. cit.*). The quote on mining is from *Merchant (op. cit.)*.

For a utopian perspective on the origins of socialism, see Martin Buber's *Paths in Utopia* (Beacon Press, Boston, 1958). For an ecological critique of Marxism, see Murray Bookchin's "Marxism as Bourgeois Sociology," in his *Toward an Ecological Society* (Black Rose Books, Montreal, 1980). Darwin's world view is lucidly explained in *Ever Since Darwin* by Stephen J. Gould (Norton, N.Y., 1977). On the rise of industrialism and the engineering establishment in the United States, see David Noble, *America by Design* (Oxford University Press, 1977). On the early history of technology, see Bookchin, Mumford, Merchant and White, all *op. cit.* The quote from Schumacher is from the Schumacher Society Newsletter (Box 76, R.D. 3, Great Barrington, Mass. 01230, Spring, 1985). The origins and social consequences of computerization are most lucidly explained in Joseph Weizenbaum's *Computer Power and Human Reason* (W.H. Freeman, New York, 1976) and in many of the references for Chapter 4.

The bioregional movement has yet to produce a thorough historical and philosophical introduction to all of its diverse activities. Kirkpatrick Sale's *Dwellers in the Land* (Sierra Club Books, San Francisco, 1985) offers a valuable earth-centered perspective on decentralist politics. Also of value is Sale's pamphlet, *Mother of Us All: An Introduction to Bioregionalism* (Schumacher Society Press, Great Barrington, Mass. 01230, 1983). *Reinhabiting a Separate Country* (Planet Drum Foundation, Box 31251, San Francisco CA 94131) is a diverse and exciting anthology of bioregional writings from Northern California, edited by Peter Berg. Another writer associated with Planet Drum, Michael Helm, has produced an anthology called *City Country Miners* (City Miner Books, Berkeley, 1982). Early work in the Portland area is described in *Knowing Home: Studies for a Possible Portland* (Rain Umbrella, Inc., 2270 NW Irving, Portland, Oregon 97210, 1981). Planet Drum's occasional newsletter, *Raise the Stakes,* offers summaries and analysis of bioregional activities around the world—their summer 1986 issue describes the Green City project in detail. An early example of American regionalism with ecological overtones is Howard W. Odum's pioneering study, *Southern Regions of the United States* (University of No. Carolina Press, Chapel Hill, 1936).

The proceedings of the first North American Bioregional Congress are available from the Bioregional Project, Box 3, Brixey, Mo. 65618, which also publishes David Haenke's small pamphlet, *Ecological Politics and Bioregionalism.*

The quotes from Gary Snyder are from an interview in *Mother Earth News* (Number 89, September, 1984). Bioregional ideas are further elaborated in his *The Old Ways* (City Lights Books, San Francisco, 1977) and in all of his recent poetry. For bioregional poetry with an East Coast flavor, see *The Gulf of Maine Reader* and other works by Gary Lawless (Blackberry, Box 687, So. Harpswell, Maine 04079).

Other quotes are from Thomas Berry, "Bioregions: The Context for Reinhabiting the Earth" and Gary Lawless, "The Bioregional Voice and the Green Movement," both included in the New England "Green Working Papers" (over 100 in all, available as individual essays from the New England Committees of Correspondence, Box 703, White River Junction, Vermont 05001), and from the "Bioregions" issue of *CoEvolution Quarterly* (No. 32, Winter, 1981). The Sale quote is from *Mother of Us All (op. cit.).*

A wealth of new bioregional publications has appeared in the last few years. Of particular note for its mixture of local reporting with reflective articles on broader ecological, historical and philosophical concerns is *The New Catalyst* (P.O. Box 99, Lillooet, British Columbia). Also of note are *Katuah, the Bioregional Journal of the Southern Appalachians* (Box 873, Cullowhee, N.C. 28723) and *Siskiyou Country* from the hills of southern Oregon and far northern California (Box 989, Cave Junction, Ore. 97523). A more complete directory of bioregional groups and publications is available from the Bioregional Project which, along with Planet Drum, has become a center for networking bioregional efforts across the continent.

2. Where Did the Green Movement Come From?

The culture and politics of the 1950's are described in Marty Jezer's *The Dark Ages* (South End Press, Boston, 1982). Also see the early poetry of Allan Ginsberg and Lawrence Ferlinghetti, published by City Lights Books, San Francisco.

The black Civil Rights movement is described in many recent histories. Particularly useful are *The Power of the People,* by R. Cooney and H. Michalowski (revised edition, New Society Publishers, Philadelphia, 1985), which places it in the context of nonviolent movements throughout American history, and *More Power Than We Know,* by Dave Dellinger (Anchor/Doubleday, 1975). The latter also describes the evolution of the anti-Vietnam War movement, as does *SDS,* a comprehensive history of the student movement (and the nationwide Students for a Democratic Society) by Kirkpatrick Sale (Viking Press, New York, 1975). The approach to community organizing that originated in Chicago is explained in a number of works by its pioneer, Saul Alinsky. A more introductory history of the 1960's can be found in the closing chapters of Howard Zinn's *People's History of the United States* (Harper Collophon, New York, 1980). A more detailed history of the movement against the Vietnam War is *Who Spoke Up?,* by Nancy Zaroulis and Gerald Sullivan (Holt, Reinhart and Winston, N.Y., 1984). Parallel events in Europe are described in Daniel and Gabriel Cohn-Bendit's *Obsolete Communism: The Left-Wing Alternative* (McGraw Hill, New York, 1968) and in some of the essays in Murray Bookchin's *Post-Scarcity Anarchism* (Ramparts Press, San Francisco, 1971, recently republished by Black Rose Books, Montreal). On the intellectual origins of the movements of the sixties, see *The Making of a Counter Culture* (Anchor, 1969) and other works by Theodore Roszak.

The rise of feminism is described in Sara Evans' *Personal Politics* (Random House, New York, 1980), with some of the important documents reprinted in *Sisterhood is Powerful,* edited by Robin Morgan (Random House, N.Y., 1970) as well as in Morgan's autobiographical work, *Going Too Far* (Random House, 1980). See also Barbara Ehrenreich's *The Hearts of Men* (Anchor/Doubleday, Garden City, N.Y. 1984). The Mary Daly quote is from *Woman of Power, op. cit.* On eco-feminism, see Ynestra King, "The Ecology of Feminism and the Feminism of Ecology" (*Harbinger,* Vol. 1, No. 2,

Fall 1983, 211 E. 10th St., New York, N.Y. 10003), and the "Feminism and Ecology" issue of *Heresies* (No. 13, 1981, P.O. Box 766, Canal St. Station, New York N.Y. 10013).

"Ecology and Revolutionary Thought" is reprinted in Bookchin's *Post-Scarcity Anarchism (op. cit.)*. The Ecology Action East manifesto is available in Bookchin's *Toward an Ecological Society* (Black Rose Books, Montreal, 1980). A more traditional environmentalist view is outlined by Barry Commoner in *The Closing Circle* (Alfred A. Knopf, New York, 1975). The origins of appropriate technology are described in *Energy for Survival* by Wilson Clark (Anchor/Doubleday, Garden City, N.Y., 1975) and in *Community Technology* by Karl Hess (Harper and Row, New York, 1979). A complete history of the anti-nuclear power movement has yet to appear in this country; one useful work focusing on the French movement is *Anti-Nuclear Protest* by Alaine Touraine (Cambridge University Press, Cambridge, England, 1982).

On the history of the Greens in West Germany, see the references cited above for the Introduction. The Hessian program is available as a Green Working Paper from the Greens in New England (see above). The 1983 national program of the Greens was published in English by Heretic Books of London, and has recently been reprinted by Inland Books, 22 Hemingway Ave., E. Haven, Connecticut 06512. The Bahro quote in this chapter is from *From Red to Green*, Petra Kelly's from *Fighting for Hope*, both *op. cit.* Most of the others are from the 1983 program. The most thorough discussion I have seen in English of the internal debates within the West German Greens, including the Fundi/Realo divisions, is Phil Hill's "The Crisis of the Greens" (*Socialist Politics* Number 4, Fall 1985, available from P.O. Box 321, Bidwell Station, Buffalo, N.Y. 14222). Chapter 5 of Carl Boggs' *Social Movements and Political Power* (Temple University Press, Philadelphia, 1986) traces the roots of the Greens in European social movements and in the unique German political culture. A good way to keep up with parliamentary developments in West Germany and across Europe is through Diana Johnstone's periodic articles in the weekly newspaper *In These Times* (1300 W. Belmont Ave., Chicago, Ill. 60657), although her coverage of the Greens is heavily biased toward those elements most engaged in the parliamentary process. The continuing role of the Greens in the West German peace movement is discussed by Peter Findlay ("Pulling Through a Difficult Patch") in the *END Journal of European Nuclear Disarmament* (No. 21, April 1986, available from 11 Goodwin St., London N1). Rudolph Bahro's letter of resignation from the Green party is reprinted in *Building the Green Movement (op. cit.).*

The vote tallies from the January 25, 1987 national election were: The Greens, 8.3% (42 Bundestag seats); Free Democratic Party, 9.1% (46 seats); Social Democratic Party, 37.0% (186 seats); Christian Democratic Union, 44.3% (223 seats). The Christian Democrats and Free Democrats constitute the current majority coalition (Source: *San Francisco Chronicle,* Jan. 26, 1987). The Christian Democrats' tally incorporates votes received by the far right-wing Christian Social Union in the state of Bavaria.

To find out how to contact the closest Green group to your own home region, write the Committees of Correspondence National Clearinghouse, P.O. Box 30208, Kansas City, Missouri 64112. The CoC newsletter is available for an annual donation of $15. Another regular Green publication of note is *Green Action* (P.O. Box 37, Tempe, Arizona 85281).

3. Ecology: The Art of Living on the Earth

On the life-sustaining qualities of the earth, see James Lovelock, *Gaia: A New Look at Life on Earth* (Oxford University Press, 1979) and *Is the Earth A Living Organism?,* the proceedings of a symposium sponsored by the National Audubon Society in 1985 (Jim Swann, ed., 1986). Also see Tobias' *Deep Ecology (op. cit.)*

On agriculture, see *Food First,* by Frances Moore Lappe and Joseph Collins (Houghton Mifflin, Boston, 1977) and *Radical Agriculture* by Richard Merrill (New York University Press, N.Y., 1976). Statistics are from *The Protection of Farmland* (National Agricultural Lands Study, Executive Summary, Regional Science Research Institute, 1980). The Cornucopia Project (Rodale Press, Emmaus, PA) has produced a series of detailed regional studies on the organic movement and its local implications. The North American Farm Alliance (Box 2502, Ames, Iowa 50010) convened the 1986 Farmers' Congress and has published much valuable material on the current crisis in American agriculture.

On the differing tendencies within environmentalism, see the *Whole Earth Review* Number 2, Spring, 1985, Kirkpatrick Sale's "The Forest for the Trees" and Paul Rauber's "With Friends Like These" (both from *Mother Jones,* Vol. 11, No. 8, November 1986), and Seth Zuckerman's "Environmentalism Turns Sixteen" (*The Nation,* Vol. 243, No. 12, October 18, 1986). On the Grand Canal idea, see *Akwesasne Notes,* Winter 1986. On wilderness and forestry issues, see the *Earth First!* newsletter (see notes to Chapter 7). For an overview of acid rain, see Robert H. Boyle, *Acid Rain* (Nick Lyons/Schocken Books, New York, 1983). The grazing issue is thoroughly explained in Lynn Jacobs' illustrated essay, "Free Our Public Lands," available from P.O. Box 2203, Cottonwood, Arizona 86326. A good summary of the rain forest issue is "Tropical Deforestation: A Global View," by Nicholas Guppy, in *Foreign Affairs,* Vol. 62, No. 4, Spring, 1984. The Rainforest Action Network can be reached at 466 Green St., Suite 300, San Francisco, Calif. 94133. Also see, Roger Lewin, "A Mass Extinction Without Asteroids," in *Science,* Vol. 234, No. 1 (October 3, 1986). The monthly *Eco-News* (North Coast Environmental Center, 879 9th St., Arcata, CA. 95521) covers developments in northern California. The Mettole Project was described by Seth Zuckerman in "The Return of the Natives" (*Not Man Apart*—the journal of Friends of the Earth—530 7th St., SE, Washington, D.C. 20003). The Mettole Restoration Council can be contacted directly at 3848 Wilder Ridge Rd., Garberville, CA 95440.

A recent overview of the science and politics of nuclear power is Dr. Rosalie Bertell's *No Immediate Danger* (The Book Publ. Co., Summertown, Tenn., 1986). A good survey of various alternative energy technologies, with an emphasis on homestead-scale applications is *Other Homes and Garbage,* by James Leckie, *et al.* (Sierra Club Books, San Francisco, Revised 1981). A classic on the issue of energy use and supply is Amory Lovins' *Soft Energy Paths* (Ballinger, Cambridge, Mass., 1977). The long-term effects of the Chernobyl accident are analyzed in a report by the Lawrence Livermore Laboratory of the University of California, described on page 1 of the *New York Times* on Sept. 23, 1986. Also see periodic articles throughout the summer of 1986 in the journals, *Science,* and *Nature* and the special June 1986 issue of *Environment.*

On the decline of public transportation see Jezer, *The Dark Ages (op. cit.).* The Worldwatch Institute (1776 Massachusetts Ave., N.W., Washington, D.C. 20036) publishes periodic issue papers on a wide variety of environmental concerns.

One recent study of hazardous waste is by the New England Congressional Institute (Report of the Hazardous Waste Management Project, Washington, D.C., 1986). To keep up-to-date on toxics issues, the publications of the Citizens' Clearinghouse on Hazardous Wastes (P.O. Box 926, Arlington, VA. 22216) are indispensible. A good introductory pamphlet is *Hazardous Waste: an Introduction*, published by the Central States Education Center (311 W. University Ave., Champaign, Ill. 61820). On the health effects of a wide variety of chemicals, see the comprehensive textbook, *Health Effects of Environmental Pollutants*, by George Waldbott (C.V. Mosby, St. Louis, 1978). For an international view, see "Many Bhopals: Technology Out of Control," by Robert Engler (*The Nation,* Vol. 240, No. 16, April 25, 1984).

For the latest findings on waste incinerators, contact the National Coalition Against Mass-Burn Incinerators (82 Judson St., Canton, N.Y. 13617). Two organizations closely involved with the development of recycling alternatives are the Institute for Local Self-Reliance (2425 18th St., N.W., Washington, D.C. 20009) and Urban Ore of Berkeley, whose activities have been documented and compiled by Materials World Publishing, 1329A Hopkins, Berkeley, Calif. 94702. On the hazards of high technology industry, see Ken Geiser, "The Chips are Falling: Health Hazards in the Microelectronics Industry" (*Science for the People,* Vol. 17, No. 1, March, 1985, available from 897 Main St., Cambridge, Mass. 02139) or Lenny Siegel, "High-Tech Pollution" (*Sierra,* November, 1984). For the most recent developments, contact the Silicon Valley Toxics Coalition, 277 W. Hedding St., San Jose, CA 95110.

The quote from Horkheimer is from "The Revolt of Nature," an essay in *The Eclipse of Reason* (reprinted by Continuum, N.Y., 1974).

4. *Social Justice and Responsibility*

On food issues in the third world, see Lappe and Collins, *Food First,* and Guppy, "Tropical Deforestation," both *op. cit.;* see also

"Africa in Crisis," a special issue of the *Bulletin of the Atomic Scientists* (September, 1985). Several Washington watchdog groups have documented the systematic diversion of federal social service funds to the military—one of the most active is the Coalition for a New Foriegn and Military Policy (712 G St., SE, Washington, D.C. 20003). The War Resisters' League (339 Lafayette St., New York, N.Y. 10012) also publishes a wealth of material on taxation and militarism.

Many small agencies in different corners of the country are experimenting with holistic approaches to community social service work. One model, focused on mental health care, is offered by Resources for Community Living, 71 East Terrace, Burlington, Vt. 05401. Judith Plant's article on feminism and bioregionalism appears in *The New Catalyst* Number 2 (*op. cit.*).

On the problems of work and employment, see Stanley Aronowitz, "The Myth of Full Employment" (*The Nation,* February 8, 1986), and a debate among several authors in the following issue. The German position is explained thoroughly in a detailed economic program, *Purpose in Work, Solidarity in Life,* which is available as a Green Working Paper from the New England Committees of Correspondence (see notes to Chapter 1, above). Mike Cooley, in *Architect or Bee* (South End Press, Boston, 1982) describes the loss of autonomy brought on by new workplace technologies. A more detailed account is offered by David Noble in *The Forces of Production* (Alfred A. Knopf, New York, 1984) and in his important three-part essay on workers' resistance to new technologies and the idea of progress, "Present Tense Technology" (*Democracy,* Vol. 3, Nos. 2, 3 and 4, Spring, Summer and Fall, 1983). The quote from Noble in this Chapter is from Part III. An incisive history of the computer industry and its social implications can be found in *The Cult of Information,* by Theodore Roszak (Pantheon, N.Y., 1986). A techno-optimist account of the prospects for workplace automation is offered by R. Ayres and S. Miller in "Industrial Robots on the Line" (*Technology Review,* May 1982).

An ecological approach to housing is outlined in *Knowing Home (op. cit.).* The concept of the Community Land Trust is described in detail in the *Community Land Trust Handbook,* available from the Institute for Community Economics, 151 Montague City Rd.,

Greenfield, Mass. 01301. On the problem of health care, see Ivan Illich, *Medical Nemesis* (Bantam Books, New York, 1977) and Samuel Epstein, *The Politics of Cancer* (Anchor/Doubleday, Garden City, N.Y., 1979). On the history of health as a social and cultural principle, see Richard Grossinger, *Planet Medicine* (Shambhala, Boulder, 1982). On education, see Ivan Illich, *Deschooling Society* (Harper and Row, New York, 1971) and Joel Spring, *A Primer of Libertarian Education* (Free Life Editions, N.Y. and Black Rose, Montreal, 1975). The works of Paolo Freire are useful in offering a more trans-cultural context.

On the effects of television on children, see Jerry Mander, *Four Arguments for the Elimination of Television* (William Morrow, New York, 1978). On the effects of computers in schools, see Sherry Turkle, *The Second Self* (Simon and Schuster, New York, 1984) and "The Computer in Education in Critical Perspective," a special issue of the *Teachers College Record* (Summer 1984). See also, David Burnham, *The Rise of the Computer State* (Random House, New York, 1983).

5. Democracy in Politics and in the Economy

On the importance of an active citizenry through history, see Murray Bookchin, "Theses on Libertarian Municipalism" (*Our Generation*, Vol. 16, No. 3, Spring, 1985, available from 3981 Boul. Ste. Laurent, Montreal). On town planning in colonial America, see Cronon, *Changes in the Land (op. cit.)*. The role of the "masterless" is suggested in Chapter 6 of *The Ecology of Freedom (op. cit.)*; on the idea of "nomadism," see G. Deleuze and F. Guattari, *A Thousand Plateaus* (University of Minnesota Press, 1987), especially Chapters 12 and 13.

On affinity groups and consensual process, see V. Coover, *et al.*, *Resource Manual for a Living Revolution* (New Society Publishers, Philadelphia, 1978, revised 1985). The quote on consensus process is from *Greenham Women Everywhere* (see notes to Chapter 6). For an enlightening sociological analysis of consensual and adversary decision-making (with an incisive analysis of a Vermont Town Meeting), see Jane Mansbridge, *Beyond Adversary Democracy* (Basic

Books, New York, 1980). On the psychology of Affinity Groups, see Joel Kovel's *Against the State of Nuclear Terror* (South End, Boston, 1984).

For analysis of the origins of the cooperative movement in Europe, see Martin Buber's *Paths in Utopia (op. cit.)*. For more recent history, *The Food Coop Handbook* (Houghton-Mifflin, Boston, 1975) is useful. Worker collectives and the coop movement in Berkeley are explored in a special issue of *Communities* magazine (No. 70, Spring 1986, available from 126 Sun St., Stelle, Ill.). A successful computerized community barter system has been established in parts of British Columbia by Landsman Associates (470 4th St., Courtenay, B.C.). A comprehensive directory of collectives and co-ops in the Pacific Coast states is available from The InterCollective, P.O. Box 5446, Berkeley, CA. 94705.

A thorough treatment of the prospects for a decentralist politics and economics is *Human Scale,* by Kirkpatrick Sale (Putnam, New York, 1982). The E.F. Schumacher Society (see notes to Chapter 1) and the Institute for Community Economics (see Chapter 4 notes) have pioneered efforts to help communities establish Community Land Trusts and Loan Funds. The Schumacher Society is also a good source of details on the SHARE program. The Mondragon System is described in detail by H. Thomas and C. Logan in *Mondragon: An Economic Analysis* (Allen and Unwin, Boston, 1982). A new quarterly journal, *Changing Work* (P.O. Box 5065, New Haven, Ct.), is published by people affiliated with the Industrial Cooperative Association, an organization actively promoting workers' cooperatives.

The use of industrial decentralization as a ploy for increasing management control is advocated in the European chapter in *The Crisis of Democracy* (New York University Press, 1975—see chapter 7). See also the works of Leopold Kohr who spoke on the subject at the New England Green conference of June, 1985.

6. Toward a World of Peace and Nonviolence

Two noteworthy anthologies of writings are *Beyond Survival,* edited by Michael Albert and Dave Dellinger (South End Press,

Boston, 1983) and *Exterminism and Cold War,* edited by the staff of *New Left Review* (Verso/New Left Books, London, 1982). On the revival of the Cold War, see Jerry Sanders, *Peddlers of Crisis* (South End Press, Boston, 1983), and anything by Noam Chomsky, especially *Turning the Tide* (South End Press, Boston, 1986) and *Towards a New Cold War* (Pantheon, 1982). The infamous *Business Week* article is from the issue of February 2, 1949. For ongoing coverage of the European peace movement, see the *END Journal* (see notes to Chapter 2).

On the economic impacts of military industry, see Seymour Melman, "Swords into Plowshares—Converting from Military to Civilian Production" (*Technology Review,* January, 1986), and a symposium on the idea of economic conversion in *Changing Work Number 2* (Winter, 1985). Mike Cooley's *Architect or Bee? (op. cit.)* describes initiatives by British aerospace workers to design products they could manufacture if their companies were to lose their military contracts. Statistics on the military budget are borrowed from an analysis of the 1986 federal budget by the War Resisters' League (see notes to Chapter 5), as summarized in a pamphlet, "Your Income Tax Dollars at Work."

For essays on nonviolence from a feminist perspective, see *Reweaving the Web of Life* (Pam Mc Allister, ed., New Society Publishers, Philadelphia, 1982) and A. Cook and G. Kirk, *Greenham Women Everywhere* (South End Press, Boston, 1983). The latter is the source of the two quotes in this chapter, both describing experiences in the winter and spring of 1983. See also notes to Chapters 2 and 5.

Alternative defense strategies are discussed in *Exterminism and Cold War (op. cit.)* and in Gene Sharp's *Making Europe Unconquerable* (Ballinger, Cambridge, 1985). The Coalition for a New Foreign and Military Policy's report, "A Few Billion for Defense," is available from 712 G St., S.E., Washington, D.C.

On the military nature of new computer technologies, see Tom Athanasiou's "Artificial Intelligence: Cleverly Disguised Politics," in T. Solomonides and L. Levidow, eds., *Compulsive Technology* (Free Association Books, London, 1985) and Paul N. Edwards' "Border

Wars: The Science and Politics of Artficial Intelligence" in *Radical America*, Vol. 19, No. 6, November 1985.

The analogies between U.S. policies in Central America and Vietnam are explored in depth in Chomsky's *Turning the Tide (op. cit.)*. For more on the Pentagon Papers and U.S. strategy in Vietnam, see Daniel Ellsberg, *Papers on the War* (Simon and Schuster, N.Y., 1972). On decentralist currents in Nicaraguan history, see Jay Moore's two-part article, "The True Story of Sandino" in the Canadian quarterly *Kick it Over* (Numbers 14 and 16, Winter 1985 and Summer 1986, available from P.O. Box 5811, Station A, Toronto, Ontario). On ecological restoration efforts in Nicaragua, see ongoing coverage in *Science for the People* magazine (897 Main St., Cambridge, Mass, 02139). On the colonization of indigenous peoples by Third World governments, see the special Summer 1986 issue of *Cultural Survival* dealing with native land rights (Vol. 10, No. 2, available from 11 Divinity Ave., Cambridge, Mass. 02138). The North American division of Oxfam is based at 513 Valencia St., Number 8, San Francisco, CA 94110. For a discussion of Green approaches to transcending the East/West split, see Bahro, *From Red to Green (op. cit.)*. The latter also discusses the pitfalls of Third World industrialization. The first of the concluding Bahro quotes is from his essay in *Exterminism and Cold War;* the second is from *Building the Green Movement* (both *op. cit.*).

"A Parable"

The prophesy of the Hopi is described in an article by Hopi interpreter Thomas Banyacya in *East-West Journal* (December 15, 1985) and in a pamphlet prepared by John Kimmey (P.O. Box 688, Arroyo Hondo, New Mexico). The traditional community of Hotevilla publishes a newsletter for international distribution, called *Techqua Ikachi* ("Land and Life"), available from Box 767, Hotevilla, Arizona. The quote from David Monongye is from a message to the Dalai Lama of Tibet dated October 1982, also distributed directly by the Hopi nation from Hotevilla. For background reading, see Peter Matthiessen's *Indian Country* (Viking Press, New York, 1984).

7. How Can We Create a Green Future?

Much of the discussion in this chapter is expanded from the author's own, "Movement or Party: Is That Really the Question?," with appendices on the Realo/Fundi split and on the Burlington situation. It is available from Green Working Papers (*op. cit.*), also reprinted in *Up from the Ashes* (No. 4, Winter 1986, P.O. Box 5811, Station A, Toronto, Ontario). The situation in New Haven was described by a delegation of Party members at the New England Green Conference of November, 1985. A complete set of 1985 election documents, news releases and clippings is available from the New Haven Green Party, 304 Alden Ave., New Haven, CT. See also, Bahro, *Building the Green Movement (op. cit.)*. An excellent summary of the different political tendencies within the Green movement can be found in David McRobert, "Green Politics in Canada" (Probe Post, 12 Madison Ave., Toronto, Ontario), and also available from Green Working Papers. Developments in Brazil are described in a *New York Times* article on the career of Green party founder Fernando Gabeira (August 5, 1986).

The Earth First! organization can be reached at P.O. Box 5871, Tucson, Arizona. Edward Abbey's *The Monkey Wrench Gang* is available in paperback from Avon Books (Philadelphia, 1975). Other publications covering ongoing ecological movements include *The New Catalyst* (see notes to Ch. 1), *Synthesis* (Box 1858, San Pedro, California 90733), *Green Letter* (P.O. Box 9242, Berkeley, California 94709), *Alternatives* (Faculty of Environmental Studies, University of Waterloo, Waterloo, Ont.) and *Kick it Over* (see notes to Chapter 6). The Greenpeace organization has a colorful newsletter in a magazine format; they can be reached at 2007 R St., NW, Washington, D.C. 20009. On issues of cultural transformation, see *In Context,* "A Quarterly of Humane Sustainable Culture," (P.O. Box 2107, Sequim, Washington 98382).

The North American chapter of *The Crisis of Democracy (op. cit.)* is by Samuel Huntington, a Harvard political scientist who designed the method of rural population control through "strategic hamlets," used by the United States in Vietnam. The study's implications are discussed in Chomsky, *Turning the Tide (op. cit.)*.

The "new paradigm" in the sciences is described in Fritjof

Capra's, *The Turning Point* (Simon and Schuster, New York, 1982). A more technical discussion with some ecological resonances can be found in Erich Jantsch, *The Self-Organizing Universe* (Pergammon Press, Oxford, 1980). See also *Is the Earth a Living Organism? (op. cit.).*

The relationship between politics, culture and ritual is explored in Starhawk's *Dreaming the Dark* (Beacon Press, Boston, 1982). See also La Chapelle (*op. cit.*) and some of the feminist sources for Chapters 2 and 6.

ACKNOWLEDGEMENTS

I would like to take this opportunity to thank all of the dozens of people without whose assistance this book would not have been possible. First, I want to thank my publisher, Robert Miles, who first proposed the idea of a Green primer, and whose unfailing support and encouragement helped me over countless difficult hurdles. I am indebted to Fred Friedman, Jim Ennis, Michael Arnowitt-Reid, Anne Genovese and Bill St. Cyr, who each read large portions of the original manuscript and offered countless invaluable suggestions.

Many others, both in Vermont and elsewhere, offered helpful editorial suggestions and moral and personal support, including Victor Manfredi, Sue Richman, Brett Portman, Jay Moore, Judy Sargent, George Longenecker, Howard Hawkins, Timi Joukowsky, John Rensenbrink, Henry Lappen, Andrea Luna, Stuart Crow Frye, Mark Howard, Annie McCleary, John Wires, Steve Soifer, Naomi Almeleh and others too numerous to mention. The late Anne Mills offered a stirring example of personal courage and strength in the face of insurmountable odds. Special thanks are also due to my close friends and co-conspirators in the Northern Spy Land Trust, the Equinox Affinity Group, the Central Vermont and Burlington Greens and the various other collectives that have provided creative outlets and personal sustenance for me over the past few years. Murray Bookchin deserves special credit for introducing me to the philosophical underpinnings of an ecological perspective and for several years of intellectual and political encouragement. All of my West Coast friends, old and new, helped make the difficult weeks of final editing a wonderfully enriching time. And, finally, thanks to Spruce Mountain, to Beaver Meadow, to the lower Lamoille River valley, the hills and lakes of the far Northeast Kingdom and the other special Vermont places that have freely offered their grandeur and spiritual nourishment time and again. They, too, express the urgency of creating a Green future.

INDEX

Abbey, Edward, 174
acid rain, 10, 68, 99
affinity group, 103-4, 105, 124
Africa, 13, 59, 80-1, 84-, 124-5, 135
agriculture, 2-3, 15, 19-20, 29-30, 57,
  60-3, 77, 81; organic, 61-3;
  urban, 63, 141
anti-nuclear movement, 43-4, 103-4;
  European, 44, 49
appropriate technology, see
  technology, alternative
arms race, 32, 115-7, 118-9, 120, 138
Australia, 134, 140

Bacon, Francis, 21-2
Bahro, Rudolph, 47, 49
Berg, Peter, 30-1
Berkeley, Calif., 2, 69, 74, 78
bicycles, 73
bioregion, 27-32, 56, 60, 64, 69, 82,
  85, 98, 137, 147
biotechnology, 78; see also, genetic
  engineering
Bookchin, Murray, 14-5, 25, 41
Boulding, Elise, 16
Brower, David, 64
Burlington, Vermont, 99, 143

California, 28, 65-6, 69-70, 73, 74,
  77-8, 104; Greens in, 2, 51, 61, 68,
  142
Canada, 50, 65-6, 86, 107, 140
capitalism, 20, 22-3, 25, 59, 68, 80,
  82, 97, 109, 125
Central America, 105, 116, 125; see
  also, El Salvador, Guatemala,
  Nicaragua
chemical pollution, 3, 48, 58, 61, 65,
  73-9, 88-9, 139
Chernobyl, 2, 59, 71
Christianity, 18-9, 102, 123; see also,
  religion
cities, 2, 16-7, 36, 58, 69, 82, 91;
  agriculture in, 63, 141; crime in, 85;
  democracy in, 99-101, recycling
  in, 74
civil disobedience, 115, 121-2
civilization, 9-11, 25, 58, 69
civil rights, 34-6
Clastres, Pierre, 12-3, 15
coalition, 139
Cold War, 47, 125, 134
Committees of Correspondence (U.S.
  Greens), 51
community, 9, 32, 55, 82-3, 93, 135,
  136, 147-50; economics, 108-12,
  141; education, 93; energy production, 71-2; organizing, 34, 35; self-
  reliance, 101-2; traditional, 14,
  149; work in, 87-8
Community Congress, 28-9, 139, 141
Community Land Trust, 63, 90-1,
  108, 141
Community Loan Fund, 91, 108, 110
competition, 4, 85, 96, 135
computers, 24, 77, 86-7, 90; in

schools, 95
consensus, see democracy, consensual
conservatives, 102, 136, 142, 150
consumerism, 14, 59, 62, 94-5, 135
cooperation, 5, 43, 85, 96, 148, 149-50
cooperatives, 71, 108-12, 140;
  food, 63, 108-9, 138
Cruise missiles, 47, 115-6, 122
culture, 25, 30, 84-5, 148-50;
  counter-, 34, 38; mass, 96, 149;
  rural, 63, 81
Czechoslovakia, 37

Daly, Mary, 39
Darwin, Charles, 23
decentralism, 4, 46, 63, 82-3, 97-9,
  112, 113, 128, 144, 146-7
deep ecology, 150
defense, 119-124
Dellinger, David, 38
democracy, 2, 3, 5, 13-4, 25, 38, 45-6,
  83, 97-113, 137, 140, 142, 146-7;
  among Greens, 102-6; consensual,
  42, 102, 104-5, 149; economic, 98,
  106-12; grass roots, 3, 50
Descartes, Rene, 21
development, economic, 13, 57, 80,
  126-7, 143; opposition to, 66-7,
  138-9, 141
dioxin, 74
disarmament, 1, 43-4, 115, 120
Dubos, Rene, 138

earth, 9, 12, 19-20, 131
Earth First!, 64, 137, 139-40
eco-feminism, 39, 85, 137
ecology, 1, 58-79, 128, 150; and
  ethics, 56, 66, 78, 136, 149-50;
  restoration, 68, 69-70; scientific,
  4, 150
economics, 4, 46, 86, 136; agricultural, 62-3, 81-2; cooperative,
  110-1, 176; international, 47, 76,
  107-8
education, 83, 88, 93-6
Egypt, 17

electoral politics, 3, 32-3, 141-6;
  in Germany, 45-49
El Salvador, 116, 118
energy, 62, 70-2, 73, 109; "crisis", 42,
  70; solar, 43, 70-1, 138
environmentalism, 40-1, 64-6, 79,
  139-40, 142-3
Europe, 59, 74, 89, 107, 117, 119-21,
  127; co-ops in, 109, 111, Greens in,
  1, 34, 115-6, 128, 134, 141, 144;
  peace movement in, 115-6; 'sixties
  in, 37, 44; transportation in, 73
evolution, 23
extremism, 116

federalism, 82-3
feminism, 2, 39, 43, 84-5, 137
food co-ops, see cooperatives
Foreman, Dave, 64
forests, 3, 18, 58, 67-9
Fourth World, 127
freedom, 5, 10, 55, 121, 148

Galileo, Galilei, 21
Gandhi, Mohandas, 121
General Motors, 72
genetic engineering, 2-3, 61, 77-8
gentrification, 99, 135
Germany, Federal Republic of (West
  Germany), 23, 89, 128; Bundestag
  (federal parliament), 45, 48, 49,
  106; Greens in, 1-2, 5, 44-9, 72, 84,
  106; Green economic program, 88-9,
  112; 'sixties in, 37, 44
Greenham Common, 104, 122-3
greenhouse effect, 10
Griffin, Susan, 21, 40
growth, economic, 3, 59, 80; see also,
  development
Guatemala, 13, 118, 127

health, 79, 82, 83, 91-3
herbicides, see pesticides
Hesse, Greens in, 45, 48
hierarchy, 4, 13, 15, 16-8, 84
Hopi, 131-3

Horkheimer, Max, 79
housing, 90-1, 112, 138
hunger, 57, 80-2, 135

incinerators, 74
Indians, American, 3, 12-3, 19, 31,
    131-3, 140
indigenous peoples, 13, 127, 131-3
industrialism, 3, 10, 22-5, 40, 43,
    49, 59, 75-6, 125, 137
Industrial Revolution, 22, 89
information society, 76, 95

Japan, 16, 87, 120, 134

Kansas City, 2, 51
Kelly, Petra, 45-6, 48, 106
Kepler, Johannes, 21
King, Martin, Luther, Jr., 35

land, indigenous rights, 127-131-2;
    ownership of, 5, 62, 140-1; see also,
    Community Land Trust
Los Angeles, 2, 43, 51
Love Canal, 75
Luddites, 89

Maine, 2, 29, 51, 142
Marx, Karl, 11, 23
Massachusetts, 78, 90, 110, 141
Mattole River, 69-70
mechanistic philosophy, 10, 21-3
media, mass, 11, 94-5, 102, 144
megamachine, 17, 72, 128
Middle Ages, 11, 19-20, 25
Middle East, 16, 84, 116, 118, 125
Midwest (U.S.), 61, 65, 74
militarism, 51, 116-8, 120, 128, 135;
    Third World, 126
mining, 3, 19, 66, 131
Mondragon, 110, 111
Mumford, Lewis, 17, 25

National Parks, 67, 140
Navajo, 3, 132
Nazism, 84, 121, 134, 148

neighborhood assemblies, 99
New England, 65, 74, 91, 98-9, 105,
    138; Greens in, 2, 51, 61; see also,
    Maine, Massachusetts, New
    Hampshire, New Haven, Conn.,
    Vermont
New Hampshire, 2, 42, 101
New Haven, Conn. Greens, 3, 84,
    141-2, 145
New Right, see conservatives
New York City, 30, 35, 51
New Zealand, 134
Nicaragua, 116, 118, 126
Noble, David, 89
nonviolence, 36, 115, 120-4, 126, 149
North American Bioregional
    Congress, 29, 32, 50
North Carolina, 142
Norway, see Scandinavia
Nuclear Free Zone, 120
nuclear power, 2, 24, 43-4, 48, 65,
    70-1, 99, 105
nuclear weapons and war, 5, 47, 99,
    118-9, 123-4

Oregon, 28, 73
Ozarks, 27, 28, 29, 32

Parks, Rosa, 35
patriarchy, 38-9, 84, 85
peace camp, 122-3
peace movement, 1, 37-8, 103-4, 115,
    127-8, 143
Pentagon, see United States, armed
    forces
Pershing II, 47, 115, 128
personal growth, 2, 26, 34, 85, 103,
    105, 137, 148-50
pesticides, 61, 67
Planet Drum Foundation, 28, 29
Plant, Judith, 86
Plato, 11, 17, 21
pollution, 10, 32, 40, 65, 68, 72, 79, 81
Portland, Oregon, 28
primitive cultures, 9-13, 15

racism, 5, 31, 84
rain forests, 10, 68-69, 81
Reagan, Ronald, 82, 146
recycling, 29, 68, 74-5, 109
reformism, 138
religion, 94, 136; see also, Christianity

Sale, Kirkpatrick, 30
Sanders, Bernard, 143-4
San Francisco, 35, 51, 66; Green City project, 2, 29-30
Scandinavia, 1, 89, 134
Schumacher, E.F., 24, 110, 127
scientific revolution, 20-3
Seabrook, N.H., 42, 44, 101
self-reliance, 4, 5, 99-101; see also, community
sexuality, 18, 34
SHARE (Self-Help Association for a Regional Economy), 110
Silicon Valley, 77
Snyder, Gary, 30
Social Democratic Party (W. Germany), 48-9
social ecology; see ecology
socialism, 22-3, 25, 59, 80, 97, 125, 143
social issues, 2, 11, 80-96
social services, 82-3
Soviet Union, 26, 37, 47, 71, 116, 117-9, 120, 121, 124, 126, 135
Spencer, Herbert, 23
spirituality, 2, 15-6, 18-9, 135-6, 149-50
student activism, 35-6, 42
Superfund, 75
sweat equity, 91

technology, 19-20, 22-6, 31, 81, 86-90, 127; alternative, 41, 43, 71-2, 74, 78, 137, 141; "high", 76, 141;

medical, 91-3; opposition to, 22, 89-90; and social change, 78
television, see media
Third World, 47, 72, 77, 80-1, 116, 124-8; Greens in, 134
Thompson, E. P., 116
Town Meeting, 98-9
toxics, see chemical pollution
transportation, 30, 72-3, 112
Trilateral Commission, 146-7
Turner, Frederick, 18-9

United States, anti-nuclear movement, 42-4; armed forces, 47, 83, 115-9, 120, 124, 126; Congress, 65, 124; economy, 26, 72, 82, 90, 117-8; elections, 141-6, government, 67, 82-3, 90, 118; Greens in, 6, 34, 50-1, 126
urban issues, see cities

Vermont, 2, 66-7, 103, 141, 142-4; see also, Burlington

war, ecological consequences, 124-5; economy, 117-8, 138; in primitive societies, 11, 16-7; opposition to, 37, 39; see also, peace movement
Washington, D. C., 41, 43, 60, 64
waste disposal, 67, 73-6
wilderness, 67, 140
women, 13, 15, 38-40, 77, 84-6; as personification of nature, 21, 40, 84; peace camp, 122-3
women's liberation, 39; see also, feminism
work, 86-90, 109
World War II, 24, 44, 86, 117, 118, 121

Yellowstone, 67

Photo by Steve Bush

Born in Brooklyn, Brian Tokar graduated from MIT (biology and physics) and Harvard (biophysics). He has worked as a consultant on environmental issues, a neurophysiologist, teacher, computer programmer, and has been an antinuclear activist, nonviolence trainer, organic vegetable gardener, community organizer, and part-time tipi dweller. He is a founding member of the New England Committees of Correspondence and the Central Vermont Greens.